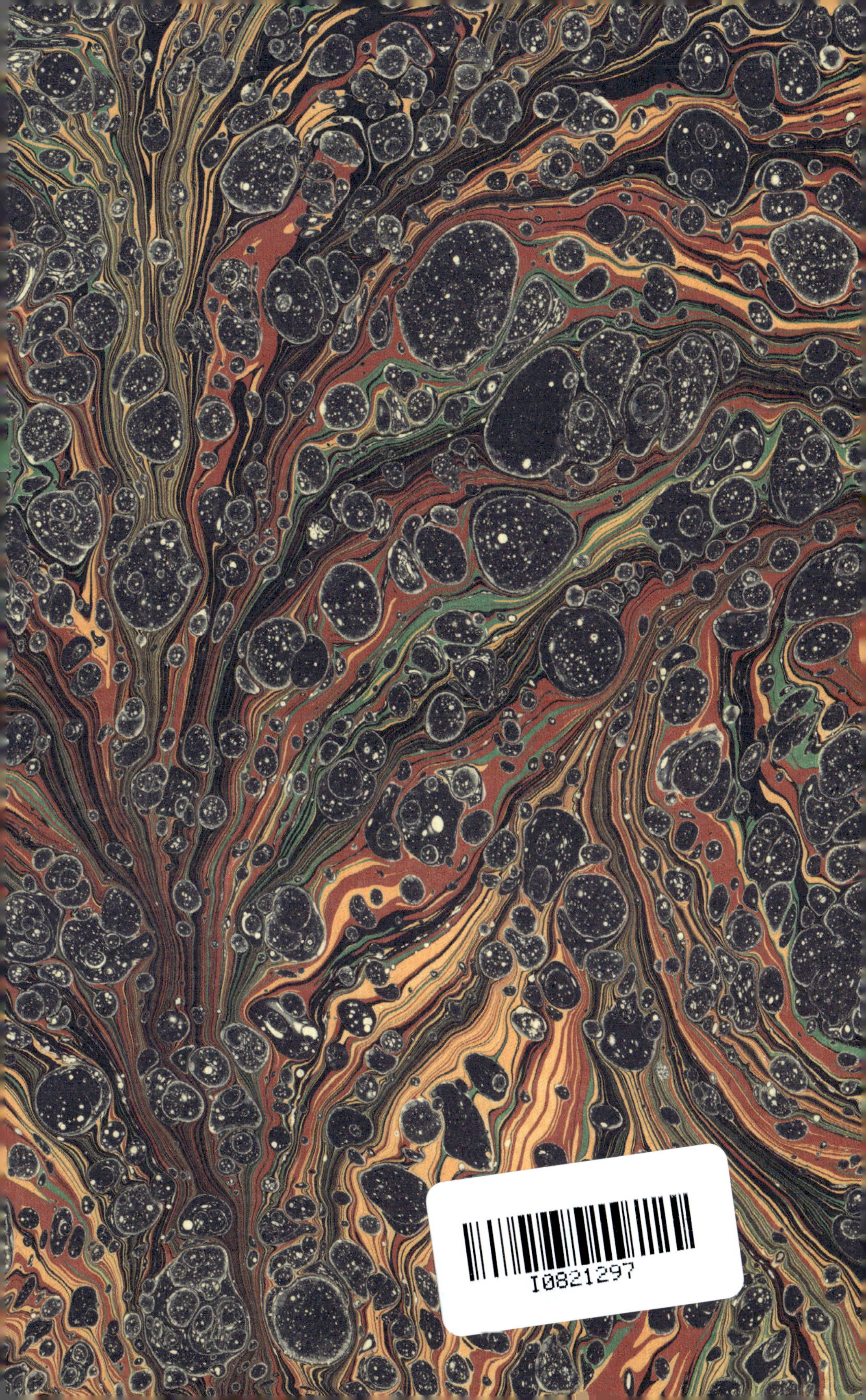
I0821297

Flak Artillery of the Legion Condor

Flak Artillery of the Legion Condor

Flak Abteilung (mot.) F/88
in the
Spanish Civil War 1936-1939

Lucas Molina Franco
& José Mª Manrique García
with R. Permuy and R. Arias
Illustrations: Ramiro Bujeiro

Schiffer Military History
Atglen, PA

Acknowledgments

We would like to express our appreciation to the following individuals and organizations for their invaluable assistance in the compilation of the photographs in this book: Josep Maria Mata, Jose Manuel Campesino, Juan Arráez, Canario Azaola Cesar O'Donnell, Revista Ejército, Francisco Marin, Carlos Franco, Constantino Lobo, Max Buch, Juan Negreira, Maria Eugenia Yagüe, Jorge Fernandez-Coppel, Heribert Garcia, and CECAF.

Book translation by Martin John Houghton

Book Design by Ian Robertson.

Library of Congress Control Number: 2008941236

Printed in China.
ISBN: 978-0-7643-3231-9

This book was originally published in Spanish under the title
Centinelas del Aire, El Grupo Antiaéreo de la Legión Cóndor en la Guerra Civil Española (1936/39) by Galland Books

Published by Schiffer Publishing Ltd.
4880 Lower Valley Road
Atglen, PA 19310
Phone: (610) 593-1777
FAX: (610) 593-2002
E-mail: Info@schifferbooks.com.
Visit our web site at: www.schifferbooks.com
Please write for a free catalog.
This book may be purchased from the publisher.
Please include $5.00 postage.
Try your bookstore first.

In Europe, Schiffer books are distributed by:
Bushwood Books
6 Marksbury Avenue
Kew Gardens
Surrey TW9 4JF, England
Phone: 44 (0) 20 8392-8585
FAX: 44 (0) 20 8392-9876
E-mail: Info@bushwoodbooks.co.uk.
Visit our website at: www.bushwoodbooks.co.uk
Free postage in the UK. Europe: air mail at cost.
Try your bookstore first.

Contents

Prologue

On 10 May 1940, the *Wehrmacht* launched its attack on the West, sweeping before it the armies of Holland, Belgium, and France, and ejecting the British Expeditionary Force from the continent. Among the many actions carried out by German forces that day, one of the most outstanding was the assault made by a handful of German paratroops on the Belgian fortress of Eben Emaël, considered the strongest fortress in the world at that time.

The fort was put out of action in a surprisingly brief space of time, which allowed the Germans to cross the River Meuse and the Albert Canal.The German ground forces given the task of supporting the paratroopers—infantry units under the command of Lieutenant Colonel Melzer and assault engineers commanded by Lieutenant Colonel Mikosch—moved with all speed to relieve them. In fact, it was Mikosch who received the surrender of Eben Emaël's garrison. Only the most erudite books on the subject mention the role played by these supporting ground forces in the operation conducted on Eben Emaël (an operation that in itself permitted the vanguard of Von Reichenau's 6th Army to enter into Belgium). However, not even these aforementioned books cite the presence in operation Eben Emaël of an anti-aircraft artillery unit called "*Flakgruppe* Aldinger," which played a vital role in consolidating the conquest of the fortress and the adjoining bridges.

It is common in many books on military history to ignore the importance of the role played by artillery in general and anti-aircraft artillery, in particular.

Therefore, those books which deal with the taking of Eben Emaël narrate in detail the actions of the brilliant paratrooper officers Witzig and Koch, and even aknowledge the parts played by Mikosch and Mezer, but it is useless to look for references to Captain Aldinger of the Anti-Aircraft Artillery.

What do these events have to do with the book I have been given the honor of writing the prologue for? Well, in the case of the Spanish Civil War the same is apparent. The majority of books dedicated to the subject do not even make reference to the existance of anti-aircraft artillery. There is, however, another connection. Captain Aldinger was not simply just any other artillery officer. Since the beginning of the 1930s he had formed part of a small but select group of German Artillery Officers who had secretly been establishing the base for the future German anti-aircraft artillery arm, something prohibited under the Treaty of Versailles. Aldinger was a member of one of the clandestine groups of anti-aircraft artillery officers that the *Reichswehr* had integrated in each of its artillery regiments for that purpose. Aldinger himself was posted to the 7th Artillery Regiment, together with eleven other officers, among whom there were men who would later become key elements in the history of the German Flak, such as Von Axthelm and Lichtenberger. This young and promising officer commanded the first contingent of anti-aircraft artillery sent to Spain, and was the first commander of the Flak Group which forms the subject of this book. However, neither he nor, what is worse, the Flak 88 Anti-Aircraft Group of the Condor Legion have ever received the recognition they deserve from military historians. Fortunately, Lucas Molina and Jose Maria Manrique have put an end to this unjust situation and undeserved obscurity that surrounds F/88.

That this unit deserves the attention of military historians (and those members of the public interested in such matters) is patently obvious when we remember that the tactical innovations that the German Flak units were to use to such effect during the Second World War had been developed by German gunners serving in Spain during the Civil War. Therefore, it becomes strikingly clear that a Spanish Civil War veteran like Aldinger was not chosen by chance for a mission of such importance as the conquest of Eben Emaël.

After such a long and undeserved silence regarding the Flak 88 Anti-Aircraft Group, I find it no small consolation to see that this situation is to be rectified by two such distinguished Spanish military historians. Given that I enjoy their friendship, the reader could be forgiven for thinking that I am exagerating, but a simple review of the list of titles published by both authors, either individually or writing together as a very effective team, would immediately demonstrate that this is not vainly inspired praise. Their book *Weapons of the Spanish Civil War* alone is sufficient to place them amongst the foremost writers covering that conflict.

Much has been written about the Spanish Civil War, and a lot of it, it must be said, is perfectly dispensable: books of personal memories which so often merely echo so many others written previously, political pamphlets thinly disguised as "history," and general histories plagued with errors. However, there seem to be very few books such as this one. Here is a book that sheds as much light as is possible on fact, in this case the history of a unit hitherto ignored. This, and there can be no other, is the role of the historian—to enable us to increase and perfect our knowledge of the past.

Carlos Caballero Jurado

Introduction

On 6 August 1936, before the official organization of what was to become known as the Condor Legion, the merchant ship *Usaramo* arrived in Seville with a small group of 20/65 mm Flak 30 light anti-aircraft guns. This coincided with the first request made by Franco to Germany (23 July), together with varied signals and aviation materiel and around one hundred German service personnel. A master armourer called Hakenholtf had been charged with training the Spanish gunners, but given the fact that he only spoke German, he was accompanied by Lieutenant Hajo Hermann, one of the Junkers Ju 52 pilots involved in the air bridge across the Straits of Gibraltar, as well as in the first bombing missions. Hermann, who went on to innovate night-fighter tactics in the Second World

The first Anti-Aircraft guns to be sent by Germany in August 1936 were a dozen 20mm Flak 30s. (C. Azaola)

War, translated Hakenholtf's instructions into French, and complimented them with his own demonstrations, such as live-fire exercises on hot air balloons.

The result of the knowledge and improvisations of the instructors, combined with the capabilities of the students, was that the anti-aircraft weapons brought from Germany were soon operational and distributed around Spain with various anti-aircraft batteries and machine gun companies. On the 25th of that same month, eight of the guns were assigned to the Nationalist Army of the North, forming the 16th Battery, 11th Light Regiment of Burgos, while the remaining twelve were assigned to the south to form three companies of the 34th "Granada" Infantry Regiment in Seville, with all of them engaged in combat the following month.

The shipments of both personnel and materiel continued at a slow rate until, almost certainly on 31 August, the first German Flak battery, complete with men and heavy equipment, arrived in Seville on board the second voyage of the *Usaramo*. Among the materiel was the Model 36 fire control system, and also four very modern artillery pieces. In command of this group was Lieutenant Aldinger, and the unit was known by his surname

Each 88/56 mm anti-aircraft battery had two 20/65 mm Flak 30s for close defense. (J. Negreira)

for some considerable time. The world famous "eighty eight," known as the "*otto con otto*" by the Spanish, had arrived on Spanish soil; the 88/56 Flak 18, which was to first see service in the Spanish Civil War. This heavy anti-aircraft battery had been valued by the Germans at 200,000 RM (Reichsmark).

Following the systematic destruction of its records at the end of the Second World War, the history of the *Luftwaffe* and of those units forming part of the Condor Legion present facets which are difficult to reconstruct concisely, and this is particularly true of the first year of its existence, making it necessary to rely on a variety of sources, including the less than precise records held by the Spanish Institute of Military History and Culture. Therefore, the task of recompiling the incomplete dispatches from April 1937 on, carried out by the historian Rafael Permuy, has been both magnificent and transcendental.

The initial deployment of the "Aldinger" battery was the defense of Seville airport (Tablada), where the first group of Spanish gunners began their training with the new equipment. As well as this, the service record of the then Captain of Artillery, Luis Micheo Casademunt, and Lieutenant Antonio Serrano Espinosa (also Artillery), shows that

On the right of the group is Captain Micheo Casademunt, the first Spanish officer to command one of the modern German 88/56 batteries sent from the Reich to aid the Nationalists in the Spanish Civil War. (C. Azaola)

the newly arrived German materiel was used to organize the first Spanish 88mm Flak 18 battery. The unit was trained by Aldinger and commanded by Micheo (together with Lieutenant Serrano, four Sergeants, eight corporals, and 40 gunners, all of whom belonged to the 3rd Light Artillery Regiment), who had received his orders to join the group at the end of September. Micheo arrived in Oropesa (Toledo) on 8 October, and the unit reached their position on the 9th; he then deployed with them to Toledo on the 27th, with the guns being used in anger for the first time that same day.

The next day saw the first attack of the war by a modern bomber, the Tupoliev SB "Katiuska." Initially, the Nationalists confused the planes with North American Martin B-10 bombers, and the Republican aircraft were able to continue on to attack Seville to the great surprise of its inhabitants.

A magnificent view of the German Model 36 fire control instrument, an electromechanical device issued to the Anti-Aircraft Batteries equipped with the 88/56 mm Krupp. (C. Azaola)

The insurgent Generals Franco and Mola, in company with Kindelan, chief of the Spanish Air Force, and Sperrle (on the right), the Condor Legion's first Commander. (Campesino, via R. Arias)

1

Origins of the Legion Condor

On 30 October, a few days after a meeting held in the Canary Islands between General Franco and Admiral Canaris, Berlin authorized the formation of a *Luftwaffe* volunteer corps, which was designated with the number 88. Other elements belonging to the Army (*Heer*) and the German Navy (*Kriegsmarine*) were also to form part of the military assistance Germany was to provide the Nationalist cause, but their organization and missions do not form part of this study. Using the excuse of "maneuvers" which were codenamed Hansa, Feuerzauber, and Rügen, 25 ships transporting men and materiel left the German ports of Stettin, Hamburg, and Swinemünde on 7 November 1936, arriving in Cadiz and Seville on the 29th. From that moment, what was to be known as the Condor Legion began to take on form. However, it should be mentioned that other small amounts of men and equipment had been arriving in Spain, and these were also integrated into the unit that was being organized.

Fundamentally composed of aviation units including fighters, bombers, and reconnaissance aircraft (which as the conflict progressed would see Spanish personnel included in their ranks), the initial composition was:

- Command and Staff (*Stab*), *Führungsstab* S/88
- Fighter Group (*Jagd*), *Jagdgruppe* J/88
- Bomber Group (*Kampf*), *Kampfgruppe* K/88
- Reconnaissance Squadron (*Aufklärung*), *Aufklärungsstaffell* A/88
- Naval Bombing and Reconnaissance Squadron (Sea) AS/88

A range of materiel belonging to the Condor Legion's Anti-Aircraft Group being loaded onto rail wagons. Here can be seen a Krupp six-wheeled truck (top), and a half-track Sd.Kfz. 7 KM 8 tractor and an 88/56 Flak 18 (below). (J. M. Mata)

- Motorized Air Signals Battalion, *Luftnachrichten Abteilung* (mot.) LN/88
- Maintenance Group and Materiel Park (Park) P/88
- Motorized Anti-Aircraft Artillery Group, Flak *Abteilung* (mot.) F/88.
- Other Service Units

By the end of that same November, the Condor Legion already had a significant anti-aircraft artillery element in the F/88 (*Flakabteilung* 88) Group, even though it was still at the stage of organizing itself and integrating the different personnel and equipment. Composition was as follows:

- Three heavy batteries (*schwere*) consisting of the 88/56 Flak 18 numbered 1st, 2nd, and 3rd.
- Two light batteries (*leichte*), the 4th and 5th, equipped with the 20/65 Flak 30.
- The 6th Searchlight Battery.
- A munitions column.

One place is as good as any to take a nap (that typically Spanish practice of the Siesta) in the heat of the sun's rays. These German gunners of F/88 are seen resting near their Flak 18 while protecting a Nationalist airfield. (Campesino, via R. Arias)

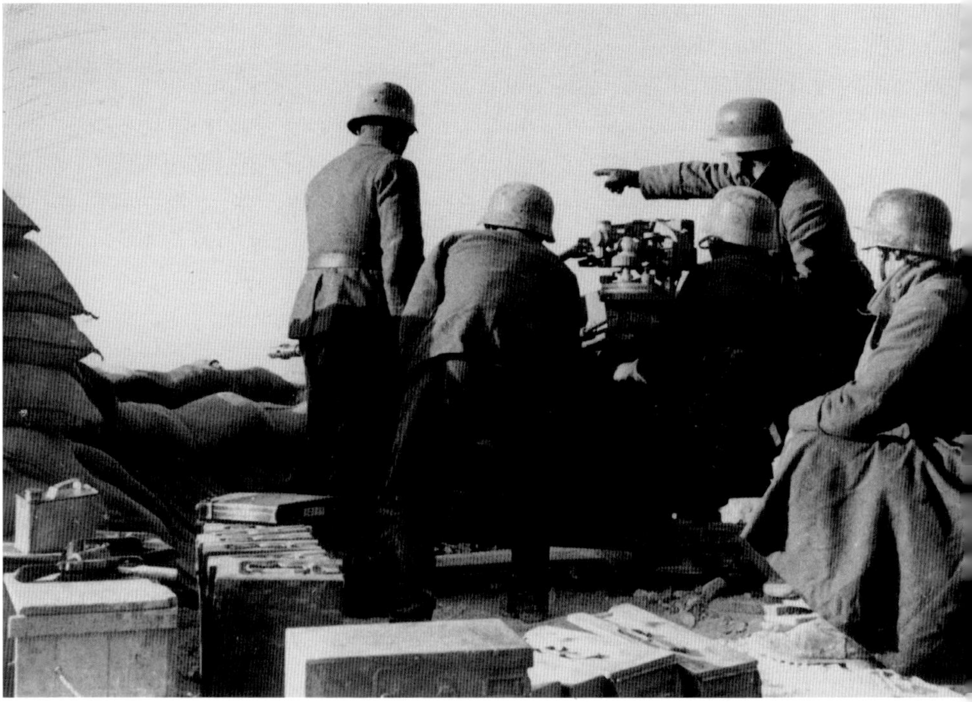

Two of the German Anti-Aircraft Battalions' batteries (the 4th and 5th) were equipped with the light 20 mm Flak 30. This was a modern weapon that was standard equipment in the *Wehrmacht* at that time, and one that was very effective against low-flying aircraft due to its rapid rate of fire. (Authors' collection and Marin)

The battery commanded by the Spanish officer Captain Micheo (although it included German personnel) was at that time engaged in operations apart from the German Group: on 2 November it was deployed at Talavera aerodrome (province of Toledo), then on the 4th it was sent to Cubas (Madrid), again to Toledo, and on the 10th to Leganes, where it shot down a Potez 540 (marked "A") to the north-east of Alcorcon, killing the international communist Primo Gibelli. The Nationalist report and some of the remaining statistics appear to confirm this, claiming that a three-engined plane crashed in flames over the Nationalist lines; the same day, between Leganes and Carabanchel, Micheo's Flak battery, the only anti-aircraft artillery unit present in the area, opened fire on a flight of Breguet XIXs. They also hit some SB Tupoliev Katiuskas, causing the death of one of the crew, Manuel Suarez. In response, the anti-aircraft battery was bombed by Republican planes.

On the 11th, six SB Tupoliev Katiuskas attacked the aerodrome near Avila, with two bombers being lost on the return flight, the official cause being listed as fog. Trautloft, a German pilot who was present on the ground during the bombing, states in his book *Als jagdflieger in Spanien* that the Nationalist Flak brought down an enemy bomber, although there is no record of which unit was responsible, but it is almost certain that it was an 88 mm Flak 18 of the recently arrived German units of the Condor Legion.

Above: Staff officers of the Condor Legion (S/88) salute the man who was to be their commander for the next year, *Luftwaffe* General Hugo Sperrle, equivalent to Brigadier General during his time in Spain. (Campesino, via R. Arias)

Micheo received orders to take his battery to Seville, doing so on 29 November, in order to complete the establishment of the 88mm batteries, handing over the materiel and staying there until 16 January, seconded to a German battery of 88s and studying the use of its equipment.

The first experience in combat of the "Aldinger" battery in Seville was described by their opponents in the book *Under the flag of Republican Spain*, in which Soviet Air Force General M. Yakushin describes how, on 7 December, eight SB Tupolievs, mostly piloted by Russians, attacked Tablada at a height of 1,200 m. Despite the fact that the description is not that explicit, it does include the action of the anti-aircraft batteries and fighters (the last mentioned maintaining a cautious distance from the Russians), relating the destruction of the aircraft piloted by Alejandro Ramos on the return flight, and that of Ananias San Juan

The heat in Spain saw the German troops issued with broad-brimmed hats, sunglasses, and lightweight clothing. These soldiers are mounting guard with their Flak 30. (Campesino, via R. Arias)

which, according to Yakushin, burnt up on the ground following a forced landing, plus the damage caused to a third that landed without its undercarriage. A fourth aircraft also had to make a forced landing but was undamaged. Salas and Madariaga only recognize the shooting down of Ramos, which they accredit to the Nationalist fighter pilot Bermudez de Castro, and the official report of the Republican Air Force also only admits to the loss of one plane, and does not consider the other two planes lost as due to enemy action. It could be said that the two planes destroyed on the ground were the result of accident or mechanical failure, although it is fair to say that the incidents could also be from the result of damage inflicted by the German 88s.

The imposing Sd.Kfz. 7 in its KM m8, 9, 10, and 11 versions was assigned to F/88 to tow the heavy 88/56 Flak 18. The example here is a KM m11 variant. (J. Mata)

2

Early Operations of the German Anti-aircraft Battalion

Little more is known of the initial actions of the Flak Group in Spain. Spanish archives only show that on 28 December, during operations on the Madrid Front, having the occasion to transfer the majority of the Condor Legion's aircraft to the aerodrome at San Fernando al de Encinas (province of Avila), Sander (the pseudonym used in Spain by General Hugo Sperrle, commander in chief of the Condor Legion) requested that a battery of 88s provide air cover, and documents from Franco's General Headquarters confirm that only four batteries of these guns were available at that time (plus the two 20 mm batteries): one in Seville, another in Salamanca (protecting Franco's HQ), and two more in the vicinity of the Front protecting ground units. By this time the necessary materiel had arrived in Seville which had been bought in order to organize the first four 88mm batteries of the Nationalist Army. Meanwhile, the Italian "legionaries" had also disembarked four batteries of 75 mm guns. Therefore, it can be appreciated that in the period from the middle to the end of 1936, anti-aircraft artillery was scarce and mostly in the process of organization.

It would seem that the problem of providing air cover for the movement of the Condor Legion's aircraft was solved by employing the 88mm battery assigned to the defense of Franco's Headquarters and transferring the battery in Seville to the Madrid Font. This left the defense of Tablada to the three fixed 47/50 Vickers, which were naval in origin and located in a position known as the "Balcony"; and of course there was whatever support could be provided by the Nationalist 88mm batteries undergoing training at the base. However, what can be ascertained for sure is that the "Aldinger" battery took part in the fighting along the La Coruña road (Madrid) in December, repulsing, together with other German batteries, several attacks by Republican aircraft, and forcing the bombers to operate at a height of 4,000m in many cases.

The first death amongst the personnel of the Condor Legion occurred on 24 December when Corporal Konrad Schfer was killed while handling a grenade on the Madrid Front. Another member of the Condor Legion killed in that same battle was Corporal Rudolf Eppert, in Villafranca (Madrid) on 14 January 1937, hit by a bullet in the chest.

In Seville, during the month of January, the formation of the four Spanish 88mm Flak 18 batteries was completed. These were then numbered 3rd to 6th in the Anti-Aircraft Artillery Group, and entered service consecutively in February and March.

At the beginning of 1937 the Condor Legions 6th battery was converted into a searchlight and heavy phono-locator battery. Some months later the munitions column was converted into a 7th battery and received reinforcements of both personnel and materiel (including vehicles). For its part, the "Aldinger" battery was incorporated into F/88 as the 8th battery.

The German *Komandogerät* 36 fire control system was a fundamental element for the 88mm Flak 18 batteries. (J. Mata)

The Republican Polikarpov I-15 fighter was first sent from the USSR and later manufactured in Spain. These were popularly known as "Chatos." (J. Arraez)

The German Anti-Aircraft Group took part in the battle of Jarama, which began 5 February 1937. According to Ramon Hidalgo Salazar in his book *German aid to Spain*, the 88s of the Anti-Aircraft Group played an important part in the rupture of the Front, employed in the role of field artillery. On the 6th, as a result of fire from a battery of the Condor Legion's 88s, Jim Allison, an American pilot flying with Lacalle's I-15 squadron, was hit by ground fire and wounded while he was strafing the gas and chemical factory in La Marañosa. On the 10th, the same squadron lost a "Chato" flown by Jose Calderon during an attack on the bridge at Pindoque. Lacalle remembered the events as follows:

"The most memorable aspect of the Battle of Jarama was the intervention of the German anti-aircraft artillery. This was to be one of the most sensational revelations during the whole air war, as it led to drastic changes in the concept and knowledge which we had had of it up until then. Their accuracy was incredibly precise, and the greatest danger for us was as we approached our targets. The first barrage almost invariably caught us at our exact altitude and just in front of the nose of my plane. I still have very clear memories of that small valley as we flew away from the bridge at Pindoque. Nobody could fly near its sides with impunity, and the need to bomb this sector was imperative. Next they sent over planes known as 'Rasantes.' In their first action the squadron became scattered all over the place. They never returned. Next came the turn of the 'Natachas,' and the same thing happened, and they received serious punishment. Finally, they sent in the fast twin-engine 'Katiuskas' with the same results. If I remember well, in their first and only mission they lost two aircraft and another was seriously damaged. In the end there was no other choice than to send back the 'Chatos.' I had to use every conceivable manoeuvre in order to avoid the deadly accuracy of this artillery. We always suffered some losses with more or less planes shot down or damaged, but we always got hit. What remains curious is that when we managed to avoid the anti-aircraft fire from below, we were hit from above by the machine guns positioned on the hillsides. I remember well that on one sortie the anti-aircraft artillery hit seven of our planes, of which two at least required more than simply patching-up. Calderon flew on my right, very close, and just as he started his bombing and strafing run, the initial barrage from the German anti-aircraft artillery had us perfectly targeted. He (Calderon) must have been hit by a shell splinter, because without making any strange movements he continued his dive until he crashed straight into the ground in that small valley on the way out from the bridge at Pindoque. One or two days after the death of Calderon, a projectile exploded just below the underside of my 'Chato,' inside the right wheel housing of the plane's undercarriage, leaving it firmly jammed. Fortunately, I was able to control my landing without flipping over, limiting myself to a small 'wheelie.'"

Personnel of I.F/88 preparing their heavy pieces for deployment. In the foreground are canvas covered ammunition baskets. (Campesino)

Above: light vehicle and personnel of I.F/88.

Chow time. The field kitchen, in this case set up in the back of a Henschel truck, was one of the battery's essential items. (Campesino)

Above: General Sperrle and his Chief of Staff, Colonel Richthofen, exchanging opinions over a map of the Northern Front. (Campesino, via R. Arias)

Below: The crew of a light 20mm Flak 30 deployed in the North taking a rest. Summer 1937. (J. Mata)

3

F/88 in the Northern Campaign

The Italian retreat in the final phase of the Battle of Guadalajara was, according to some German sources, contained with the aid of attacks by the He 51s and the efforts of the batteries constituting F/88. Following the results of that unfortunate battle, which paralyzed the attack on Madrid, as well as the reorganization of the Condor Legion, on 29 March 1937 the German unit was transferred north in order to participate in the Vizcaya offensive. One of its 88mm batteries and half of a 20mm battery deployed at the aerodrome

An artillery observation post somewhere on the Northern Front. Such posts were never without their rangefinder scope and situation maps. (Capesino)

at Villafria (bombers), in Burgos, while at the same time another heavy battery (the 1st) and half a light battery did the same at the aerodrome at Vitoria (fighters and attack aircraft), protecting the Condor Legion's aircraft. The other two 88/56 heavy batteries and the remaining light battery (the 5th), which appear to have been brought together under the command of Major Kathmann, moved forward to the front line among the deployed field artillery where, apart from providing air cover, they also engaged in bombarding enemy fortifications and counter-battery fire.

The Nationalist offensive against Ochandiano began on 31 March. By 1 April the 3rd Navarra Brigade had occupied Hill 813 with the support of the two German 88mm batteries. The next day the German contingent came under well directed enemy counter-battery fire. A premature explosion occurred in a 3rd Battery piece in Urbina, causing the immediate death of Corporal Karl Rettenmaier and seriously wounding Sergeant Emil Creutz (who died in hospital in Vitoria on the 4th), as well as Corporals Johann Fischer (died on the 20th), Roman Bigalk, and Xaver Fanderl. On the 3rd it was the Germans' turn to open a counter-battery barrage, firing 24 shells at a Republican battery.

This was to be the form of combat that characterised the operations of F/88 in Vizcaya: few anti-aircraft actions and their habitual employment in the ground fighting, with a great demand for the mobility which the powerful trucks and half-tracks gave them along the mountain roads and tracks and, of course, their accuracy when used against enemy fortifications, often managing to fire their shells directly into the embrasures. To this can be added the fact that both the heavy 88mm Flak 18s and the 20mm Flak 30s took part in the attack on the "Iron Belt," even deploying in the front line, where they suffered several casualties caused by "friendly" fire from misplaced Italian bombs.

All of this left a magnificent impression on the Commander of Artillery in the northern Army, Colonel Carlos Martinez de Campos, even though they had not had occasion to shoot down any enemy aircraft. During this campaign an attack was launched against Segovia,

Left and above: all of Flak *Abteilung* (mot) batteries had their own cover against low flying aircraft and infantry attack. In this example it is a Maxim MG 08 7.92 x 57mm mounted on an anti-aircraft tripod. (Campesino)

later known as the Battle of the Farm, between 30 May and 2 June. Although the German Anti-Aircraft Group did not participate, the 3rd Spanish Battery (equipped with 88s) was engaged, deploying rapidly to Quitapesares to protect the Nationalist reinforcements.

In the wake of these battles there is little more to recount on the actions of F/88 at this time, apart from on 4 July when the the two heavy batteries deployed in Burgos reacted to a formation of enemy bombers that were out of range and, consequently, were able to continue undamaged.

F/88 in Vizcaya, Spring 1937

Days	**Most important actions and accidents**
7-APR	Various "Chatos" driven off.
15-APR	Death of *unteroffizier* Schweckendiek (4.F/88) in Vergara from shrapnel and *Gefreiter* Erich Zschetzscing (2.F/88) from a hand grenade.
22 and 23-APR	Supporting 1st Navarra Brigade in Udala, Artecalle, and Memaya.
25 and 28-APR	Supporting the attack at Durango.
7/8/10 MAY	Ditto west of Guernica .
13/14-MAY	Ditto west of Durango.
15/16-MAY	Ditto north and west of Amorebieta.
19-MAY	Ditto in Munguía.
22-MAY	Five "Chatos" driven off. Fire against ground targets.
23-MAY	Supporting 2nd Navarra Brigade west of Durango.
26-MAY	The Orduña-Amurrio road. Harrassing fire provided north of Orduña. Supporting 3rd Navarra Brigade.
12-JUN	Breakthrough of the "iron belt: "Shooting with great results" within the general preparation for the attack, later giving support to the breaching of the enemy lines. Italian bombing of 5.F/88 in Munguía. Death of *Unteroffiziers* Felix Claus, Richard Steeg, and Frederich Bauer, as well as *Gefreiter* Martin Hoffman. Material damage.
13-JUN	5th Batt (20 mm) in the front line.
14 to18 JUN	Attacks made with two batteries of 88s and one of 20 mm.
19-JUN	Fall of Bilbao.

Left: the German Mod. 35 helmet, pistol holster, and boots are distinctive elements that clearly denote the nationality of this member of F/88. (Campesino)

Next page: personnel manoeuvring a 88/56 gun. (Campesino)

4

The Battle of Brunete and the Fall of Northern Spain

The Battle of Brunete began on the night of the 5/6 July, and saw the participation of four hundred Republican aircraft and a large concentration of units from the formation known as the Special Defence Against Aircraft (DECA in Spanish), among which were five heavy batteries operating in unison with the Republic's Army of Manoeuvre, which totalled 80,000 men supported by 130 tanks and 40 self-propelled heavy machine-guns. Following the initial intervention of some air units and, once the full significance of the attack had been appreciated, the offensive against Santander was halted on the 8th, and the entire

Left and above: personnel and materiel of I.F/88 on the Madrid Front. Summer 1937 (Campesino)

Above: Transporting heavy 88/56 Flak 18. (Campesino)

Right: Command post of the 1st Battery of the German anti-aircraft contingent in Elorrio (Vizcaya) in May 1937. In the photograph, an NCO of F/88. (Campesino)

Condor Legion made a rapid change of deployment. At the end of this redeployment the Condor Legion had established a base at Matacan aerodrome (Salamanca) for its bombers, and at Villa del Prado (Madrid), Escalona del Prado (Segovia), and Avila for its fighters. It is highly likely that its anti-aircraft artillery continued with the same mode of deployment, with half protecting the aerodromes and the other half cooperating with ground forces.

German reports take up the activity of their anti-aircraft units from the 11th, the day on which a battery fired at ground targets near Chapineria, while also claiming to have shot down two Polikarpov I-15s. The shooting down of the planes is confirmed in documents in the Spanish Air Force archives and corresponds with those that Patrick Laureau cites as forced landings due to anti-aircraft fire on the 10th. On the 12th a SB Tupoliev was shot down over Sevilla la Nueva, as was confirmed by the Republican newspaper "ABC." The following day 2.F/88 shot down a Natacha R-Z attack plane belonging to the 1st Squadron of the 30th Group—piloted by Luis de Frutos—in the vicinity of Boadilla del Monte.

The anti-aircraft engagements continued in this way as well as the participation in the ground fighting, especially on the 14th and 18th, including counter-battery fire on the 18th. On the 19th the guns of F/88 fired against both ground and air targets. On the 24th there was an accidental explosion at the battery defending the base at Avila which resulted in five gunners being wounded, one seriously. The next day there was more activity against aerial targets, as well as fire support for the infantry against the retreating enemy.

The final actions took place on the 27th, when enemy aircraft bombed Avila and Salamanca, although with little effect thanks to the anti-aircraft artillery. Two aircraft were hit at Salamanca, although there were no Nationalist claims of any planes shot down that day. However, a Katiuska pilot, Francisco Lopez Dominguez, alias "el guardia," recounted to the author Jose Luis Infiesta that the previous day the patrols of Pereira and Lopez had attacked the 88mm batteries on the Brunete Front and had lost three bombers. Were these the planes brought down by the Flak, and were never claimed due to confusion as a consequence of attacks on other days? Or, did they correspond to the actions of the 5th and 6th Nationalist batteries deployed on the battlefield (although these made no claims either)?

Komandogerat 36 fire control system of I.F/88 in the vicinity of Bermeo (Vizcaya). (Campesino)

Above: The F/88 command building in Vitoria. In the right foreground is a Sd.Kfz 7 KM m 8, and in the background are several of the Group's six-wheel trucks. (Campesino)

Following page: The work of a master armourer on a 88/56. The breech has been dismounted, and the tube is at its maximum point of recoil, allowing the German technicians to carry out battlefield repairs. (Campesino)

Once that battle had been completed, and in order to continue the Santander campaign, the Condor Legion was transferred to the Aguilar de Campoo (Palencia) sector on 29 July, establishing bases at the aerodromes of Alar del Rey -Nogales- and Herrera de Pisuerga -Calahorra de Boedo- (fighters), and with the Junkers Ju 52s in Burgos. The enemy made various bombing attacks on this city on 31 July and again on 1, 12 (110 shells fired at six Tupoliev SBs), and 13 August (one battery fired 132 rounds at five Tupoliev SBs). The attacks were stopped by the German Flak, which had deployed a second heavy battery there. Meanwhile on the 5th, at the Front, four Polikarpov I-16s had strafed the crews of various 20mm cannons, causing several casualties.

Ground operations began for the German gunners on 14 and 15 August, when they were in action north of Aguilar de Campoo (on the 15th Corporal Otto Hermann was accidentally killed by a Spanish soldier in Alar del Rey). On the 17th two heavy batteries opened fire on four I-16s, as well as engaging ground targets.

General Martinez de Campos was almost certainly referring to 18 August when he cited that the 88s were employed for the first time as anti-tank guns before Reinosa, as on that day it was reported that they had fired at ground targets to the north of that town. However, the first use of the German 88 in this role, which was to bring it so much fame during World War II, had actually occurred earlier. The honour went to the Flak 88s of the Spanish batteries. On 11 May of that same year, the 3rd Battery of the Nationalist Anti-Aircraft Group had opened fire on two Soviet T-26B tanks in the grounds of "La Buena Vista," south of Toledo. The Germans, for their part, were to again use their 88s in the anti-tank role in the Asturias campaign, also according to General Martinez de Campos.

The campaign in Santander continued with ground fire on 21 August, and there was also another accidental explosion amongst the gun positions. The next three days saw the by now familiar deployment of two heavy batteries with half a light battery, both north of Barcena and at Torrelavega, a town which fell on the 24th to the 1st Navarra Brigade. This closed the trap around Santander, which was taken by the Nationalists on 26 August 1937.

Hauptmann Noffke, commander of the first F/88 battery in September 1937. (Campesino)

As the Condor Legion's presence was not required to halt the Republican offensive against Belchite, they continued with their operations in Asturias at the beginning of September, advancing from west to east. Specifically, on the 2nd 4.F/88 claimed it had shot down a Polikarpov I-16, as well as participating in the ground fighting. The rest of the actions are summarized in the following table, where it can be observed that during the course of two months ground fire was employed on no less than 37 days, while anti-aircraft fire was only used on four days, with three aircraft claimed as shot down.

The following table details all the actions on the Northern Front in order to highlight the number of times F/88 was used in the role of field artillery, as well as the variety of missions and targets involved.

F/88 in Santander and Asturias, Autumn 1937

Day	Participation of F/88 in ground-air combat
4-SEPT	Ground fighting in the area of Llanes and west of Unquera.
6-SEPT	Anti-aircraft fire against four "Chatos" and three "Moscas" in Llanes.
7-SEPT	Two batteries providing ground fire in Llanes/Barro/Priego.
9-SEPT	Diverse ground fighting.
11-SEPT	*Gefreiter* Georg Kohlheim injured by shrapnel in Llanes.
12-SEPT	Supporting the 3rd Navarra Brigade east of Mazuco. Anti-battery fire with aerial observation. 695 shells fired.
13-SEPT	3.F/88 brings down a "Chato" in Llanes; the pilot, Ramon Ribes, taken prisoner. Ground support given to 1st Navarra Brigade (two batteries).
14-SEPT	Diverse ground support.
16-SEPT	Two heavy batteries support 1st and 6th Navarra Brigades.
17-SEPT	Fire against ground targets.
20-SEPT	8.F/88 attacks Peña Blanca.
21-SEPT	Fire against ground targets.
22-SEPT	Fire against ground targets with light 20 mm cannons.
23-SEPT	Two heavy batteries support the 1st Navarra Brigade in Benzua, and the 20mm cannons are deployed in the front line.
24 -SEPT	Ground fire from two batteries of 88s and one of 20mm cannons in Nueva/Benzua (supporting the 4th Navarra Brigade).
25-SEPT	Ground fire in Ibeo. Supporting the 6th Navarra Brigade.
26-SEPT	3.F/88 shoot down a "Rata" en Ibeo; the loss is not mentioned in Republican dispatches. Ground fire in Ibeo, Tejedo, and north of Tejedo in support of the 4th, 5th, and 6th Navarra Brigades. Harrassing fire against Ribadesella. The 20mm cannons employed against enemy infantry.
27-SEPT	Harassing fire against Ribadesella bridge and support given to the 4th Navarra Brigade.
28-SEPT	Two troops of light guns support 5th Navarra Brigade. A premature explosion in a Spanish battery results in the wounding of Sergeant P. Schalm (a German instructor).

29-SEPT	Four 88mm pieces and four 20mm cannon support the advance of the 1st, 5th, and 6th Navarra Brigades.
30-SEPT	20 mm guns used to destroy railroad infrastructure (rails and trains). Two batteries support the 5th Navarra Brigade.
1-OCT	4th and 6th Navarra Brigades supported by one heavy and two light batteries. Another heavy battery lays down harassing fire.
2/4-OCT	Bombardment of roads and villages west of Ribadesella.
5-OCT	A heavy battery accompanying the 5th Navarra Brigade lays down harassing fire west of Ribadesella.
6-OCT	Same as the previous day. Two men injured by premature explosion.
7-OCT	Heavy batteries bombard heights west of Igneto. The light batteries attack machine gun nests.
8-OCT	Two heavy batteries with the 4th and 5th Navarra Brigades respectively (together with two troops from light batteries).
9-OCT	Two heavy batteries with the 4th Navarra Brigade, two troops of light guns with the 5th, and another heavy battery attack west of Ribadesella.
10-OCT	Two heavy batteries and several light troops support the advance of 1st, 5th, and 6th Brigades .
11/12-OCT	A heavy battery and two troops of 20mm cannons support the 5th Navarra Brigade. Another heavy battery attacks west of Ribadesella.
13-OCT	Supporting the 1st and 5th Navarra Brigades advancing from the west and northwest of Cangas de Onis.
14-OCT	Two batteries of 88s support the 1st Navarra Brigade and another two attack to the west of Ribadesella. The 20mm cannons also in action.
15-OCT	Similar to the previous day, but with only one heavy battery engaged to the west of Ribadesella.
16-OCT	Support given to both the 1st and 4th Navarra brigades.
17-OCT	Similar to the previous day, with some support given to 5th Navarra Brigade.
19-OCT	Cooperating in attack of 1st, 4th, 5th, and 6th Navarra Brigades.
20-OCT	Three batteries of 88s and the batteries of 20mm cannons supporting ground operations.
22-OCT	Resting. Half the Group located in Santander and the other half in Leon.

5

Reorganization and the Offensives in Aragon

With the end of the war in the North there came a month of rest and reorganization. This same period saw the first change in the command of the Condor Legion, as Hugo Sperrle was relieved by Helmuth Volkmann on 30 October, and Plocher a von Richthofen took over as Chief of Staff. During this reorganisation the Flak Group received two troops of 37mm (37/57) Flak 18 automatic cannons, one troop being assigned to each of the light batteries.

Interesting photograph showing one of F/88's light batteries with its corresponding means of transport. Apart from the 2cm Flak 30s, in the foreground there is also one of the 3.7cm Flak 18s incorporated into the batteries. (J. Mata)

Brigadier General Helmuth Volkmann, appointed commander of the Condor Legion in October 1937; he succeeded Hugo Sperrle. (Buch, via R. Arias)

Colonel Hermann Plocher was appointed as Volkmann's Chief of Staff; he was later promoted to General. (Buch, via R. Arias)

A moment for rest and music from an accordian. These soldiers of F/88 are seen taking a break in the shade of a tree, their transport close by. (Campesino)

Ramon Hidalgo Salazar states that due to the numerous changes in deployment that the Condor Legion had been subjected to, the Staff decided to organize a locomotive and twelve railroad wagons in order to accomodate the Legion's Staff and Headquarters, as well as all the necessary elements for living and maintenance, to facilitate movement from one Front to another. Salazar also specifies that the train had emplacements with 88s for possible defence, which also means that the train included the necessary platform wagons for the transport of a battery of this type.

On 29 November S/88 was transferred to Almazan (Soria), and at least one heavy battery went to Zaragoza; on 10 December, two months after the famous surprise attack on the Sanjurjo aerodrome, the battery fired at three Polikarpov I-16s flying over the city. Other batteries were deployed to Almazan and Burgo de Osma.

Then came the decision to prepare a large scale offensive on Madrid, with the idea of carrying out a similar manoeuvre to the one at the earlier Batalla de Guadalajara, but with more men and materiel. The result of the preparation for this new offensive was that on 15 December the Popular Army launched its own offensive in Teruel, three days before the Nationalist offensive was due to commence.

The participation of the Condor Legion and, by extention, F/88 during the initial phase of the Battle of Teruel is rather vague. Hidalgo Salazar cites varias possible causes for this, pointing to the fact that Volkmann was opposed to the cancellation of the Madrid offensive and tried to increase German influence in the conduct of the war, while back in Germany important events were taking place, such as the destitution of the war minister, General Blomberg, which occurred on 4 February 1938, as well as the preparations for the annexation of Austria (completed on 13 March), with the resulting risk of war. It was precisely this last event which provided Volkmann an excuse to put forward the idea of withdrawing the Condor Legion from the war.

Whatever the case, the dispatches from F/88 during the last twenty days of December 1937 reported nothing of note, despite the death of *Unteroffizier* Ludwig Florezack on 28 December as a consequence of the bombing of Bezas (Teruel). The Nationalist counter-offensive began two days later, and *Gefreiter* Heinrich Finger was also killed in action.

Any place would do when it came to taking a rest. In this case, the running board of a half-track artillery tractor Sd.Kfz 7 (J. Mata)

The Condor Legion was assigned to the Army Corps deployed to the north of Turia (later designated the Army Corp of Galicia, under General Aranda), setting up its bases at the aerodromes of Valenzuela, Sanjurjo, Gallur, and Calamocha (all in Zaragoza and its province), and Alfaro (Navarra), which in turn dictated the deployment of its anti-aircraft units.

From 4 to the 12 January 1938, dispatches from F/88 began to reflect the fighting being experienced by the German anti-aircraft gunners, both with aircraft and enemy ground forces. These actions are summarized in the following table of events.

The Teruel Offensive, January 1938

Day	Participation of F/88 in air-ground combat.
4-JAN	Ground targets engaged north and west of Teruel.
5-JAN	Similar to the previous day. 12 I-15s put out of action. Death of *Unteroffizier* Arno Lampe by an exploding shell.
6-JAN	8.F/88 forced the retreat of several Tupoliev SBs in Calamocha Anti-air and anti-tank targets engaged at the front
7-JAN	Surrender of Teruel. 8.F/88 (Calamocha) engages 18 Tupoliev SBs, suffering casualties from the bombs. 1.F/88 and 6.F/88 bring down two "Natachas" of the 1st/30 at the front (San Blas/Caude).
8-JAN	1.F/88 engages 15 Polikarpov I-15s in San Blas-Teruel.
9-JAN	An attack by guerillas on a truck (north of Bronchales). Three men wounded.
10-JAN	Enemy counter-battery fire (approx 500 rounds fired) against 1.F/88. 5.F/88 near San Blas.
11-JAN	6.F/88 and 8.F/88 deployed in Calamocha, fire on two I-16s. 5.F/88 engages three I-16s in San Blas.
12-JAN	22 Tupoliev SBs destroyed in Calamocha: 8.F/88 claims one and another two claimed between 6.F/88 and 8.F/88. The reports conserved in the Air Force Historical Archives (IHCA) confirm two sure and one probable.

Two examples showing the enormous size of the tractors for towing the heavy 88 mm Flak 18 (above) and the *Komandogerat* 36 fire control system. These were manufactured in Germany by Krauss-Maffei and could tow up to eight tons of equipment. (J. Mata y Campesino)

The Nationalist offensive was renewed on the 17th, and the battle of Alfambra took place on 5, 6, and 7 February, with Teruel retaken on the 22nd. Curiously enough, throughout these dates the Condor Legion's anti-aircraft group hardly engaged in combat. On 28 January 2.F/88 opened fire on a reconnaissance plane which was flying over the base at El Burgo de Osma (Soria). On 5 and 6 February a heavy battery supported the attack to the east of Rubielos, and between the 19th and the 22nd a heavy battery cooperated with the infantry in attacks on the positions at "El Chopo," Mansueto, and Valdecebro, all northwest of Teruel. It is worth mentioning that on 20 February a troop of automatic 37/57 Flak 18s claimed to have brought down a Polikarpov I-15 to the west of Teruel, which was later recorded in the Air Force's monthly journal.

The Aragon offensive, which the Nationalist Army began on 9 March 1938, was planned to be developed in three phases or stages. The first phase was an attack south of the Ebro towards Vivel del Rio, a town which had been on the northern limit of operations during the Battle of Alfambra, towards the line Calanda-Caspe (Guadalope river). The second phase was an attack to the north of the Ebro towards the Cinca River. The third and final phase was an advance to the sea south of the Ebro.

In the first phase the Condor Legion was to support the Moroccan Army Corps commanded by General Yague (which consisted of the 5th Navarra, 13th, 15th, and 150th Divisions). South of Yague's Corps was the Liaison Detachment (General Garcia Valiño), the Italian "CTV," the Galicia Army Corps (Aranda), and the Castille Army Corps (Varela). General Yagüe was given the mission of recovering Belchite and occupying Caspe, supported by the Tank Group (Panzer Is and T-26s).

2.F/88 and 6.F/88 were active on the 9th; equipped with 88s, they were initially assigned to the defence of the Alfaro aerodrome, while the 4.F/88 with their 20mm cannons were given the Huerva sector in order to support the breakthrough at Aguillon and Fuendetodos. The same deployment was maintained on the 10th (occupation of Belchite) and 11th, with the success of the attack by the reserve (105th Division) due in no small part to the support of F/88. At that time much use was made of mobile units mounted in trucks, which followed the tanks and Flak detachments; with their flanks guarded by ground-attack aircraft they were able to penetrate deep behind enemy lines. This was how 6.F/88 operated east of Belchite against an enemy anti-aircraft battery composed of three Russian 7.62/55s crewed by French gunners from the International Brigades, who ended up being taken into captivity.

The 88/56s took part in the spectacular advance (36 kilometers) made by the 5th Navarra Division on 12 March between Belchite and Escatron, the last town offering significant resistance that required the intervention of the tanks and ground-attack planes, and the Popular Army's 44th Division was overrun. In order to continue the advance of the motorized columns, which were being protected by the 88s, it was nessecary to change the positions of the anti-aircraft guns three times during that day. In fact, F/88 were rehearsing what in effect was to be later known as *blitzkrieg*, or "lightning war."

Left and previous page: details of the portentous German 88-mm cannon Flak 18.

Above: various members of F/88 observe a captured Russian 76.2 mm Model 1931 anti-aircraft gun. (C. Azaola)

The next day four German batteries were involved in supporting the Spanish infantry. On the 14th, two batteries were active giving support in the Caspe sector against the Escatron-Caspe road, contributing to the halting of the enemy counter-attack. On the same day the contraversial General Volkmann, faced with what he considered the lack of decision on the part of General Yagüe in continuing the attacks in depth, gave the order to his command not to collaborate in operations. Support against air and ground targets continued on the 15th and 16th, and the enemy counter-attacks at Caspe were halted and the town taken on the 17th.

As General Vicente Rojo affirmed:

"...in only four days Yagüe's offensive totally disolved the front and sowed panic amongst the Republican forces."

Those semi-motorized divisions equipped with American Ford trucks, preceeded by the German/Spanish tanks and the German Flak, and supported by the aircraft of the Condor

Previous page: A heavy 88 mm Flak 18 can be seen in the foreground, emplaced for the defence of an aerodrome, and behind it is a Junkers Ju 52. (Campesino, via R. Arias)

Above: one of the 37mm Flak 18s belonging to 5.F/88 engaging ground targets.(Authors)

Below: unofficial emblem of 8.F/88. (Via the authors)

Legion, demonstrated that this new concept in warfare was viable. Almost one hundred km were covered in only eight days, something completely unimaginable until then.

At this time command of the German Anti-Aircraft Group was held by *Oberstleutnant* Lichtenberger, with the following as battery commanders:

1st Heavy Battery (88/56 Flak 18s) - *Hauptmann* Heitzmann.
2nd Heavy Battery (88/56 Flak 18s) - *Hauptmann* Hein.
3rd Heavy Battery (88/56 Flak 18s) - *Hauptmann* Hahn.
4th Light Battery (20/65 Flak 30s), *Hauptmann* Wilke.
5th Light Battery (20/65 Flak 30s), *Hauptmann* Wolfram.
6th Searchlight Battery, *Hauptmann* Wantig.
7th Munitions Battery, *Leutnant* Schlinder.
8th Heavy Battery (88/56 Flak 18s) - *Oberleutnant* Franz.

These were soon joined by the 9th Battery of 88/56 Flak 18s under German command and with mixed Spanish/German personnel.

Franco decided on 15 March to change his initial plan and exploit the success gained by trying to reach the Mediterranean Sea, and gave the attack north of the Ebro second priority. Therefore, the Morroccan Army Corps had to change its disposition, dragging along the Condor Legion with it in the process, which meant moving to new aerodromes at Escatron and Alcañiz, together with those at Sanjurjo and Alfaro.

It was during the redeployment for the second phase of the offensive that on 22 March a battery of 88s supported the attack of the Navarra Army Corps (commanded by General Solchaga) against Lierta, while another supported the Aragon Army Corps (under Moscardo) against Tardienta. A troop of 37/57 Flak18s, assigned to the light battery also taking part in the attack, achieved excellent results against machine-gun nests in Quarte. It was also on this same day that Yagüe crossed the Ebro by surprise between Pina and Quinto. The following day one heavy and one light battery of F/88 gave support to the Aragon Army Corps.

Generals Yagüe and Kindelan seen here during the Battle of the Ebro. (Via the authors)

In front of this 88/56 gun are the ubiquitous ammunition baskets. (Campesino, via R. Arias)

This became the trend with the attacks, of special note being the motorized advances of the 5th, 13th, and 150th Divisions (40km on the 25th and another 45km the following day), the conquest on the 27th of Barbastro and Boltaña (involving both the Navarra and Aragon Army Corps), and Yagüe's occupation of Lerida (2nd April). Two or three heavy batteries plus a light battery supported the advances and no enemy aircraft were engaged. However, on the 4th 8.F/88 shot down two Polikarpov I-16s, while four heavy batteries were providing ground support, as well as a light battery—a trend maintained until the 7th, the date Tremp was captured. A Republican General wrote "our CE XII Army Corps was pulverized...."

By this time the Condor Legion had received orders not to approach within fifty kilometers of the French frontier in order to avoid French suspicion (there had been debates in the French parliament over the possibility of occupying Catalonia), given the complicated international situation regarding the Sudetenland. It must be remembered that *Anchluss*, the union of Austria and Germany, had occurred on 13 March, and Hitler's next objective was the recovery of the former German territories that had been given to Czeckoslovakia following the Versailles Treaty.

From 10 April F/88 renewed operations supporting the Navarra Army Corps with two heavy and one light battery. On the same day the Group claimed to have shot down a Tupoliev SB north of Balaguer, according to statistics from Franco's General Headquarters. The following day the northern wing of the Nationalist offensive, formed by the Navarra, Aragon, and Moroccan Army Corps, completed their occupation of the line Sort-Tremp-Balaguer (both bypassed)-Lerida-Mequinenza, generally defined by the rivers Noguera Pallaresa and Segre, and detained their advance and carried out operations to mop up the remaining pockets of resistance. Meanwhile, the southern wing of the Nationalist offensive had ground to a halt.

While the forces occupied with relaunching the 3rd phase of the Nationalist offensive south of the Ebro were reinforced, the Condor Legion's Anti-Aircraft Group continued to support the forces located north of the river. On the 11th they were active in Conques,

Welding a piece onto the carriage of a Flak 18. The F/88 workshop was always ready to help more than one battery out of a jam. (Campesino)

and on the 12th in Belcaire, Linola, Balaguer (anti-aircraft fire), and Boltaña. On the 13th four heavy batteries and one light battery were containing enemy counter-attacks. On the 14th the heavy batteries were again in action in support of the Italian "CTV" on the south bank of the Ebro. On the 15th, which was Good Friday, the Nationalist Army reached the Mediterranean at Vinaroz, with three heavy and one light anti-aircraft batteries involved. Finally, on the 17th and 18th one heavy battery was in action. The 18th saw the end of the Battle for Aragon, with all enemy resistance ended between Vinaroz and the Ebro.

Above: Various officers from the HQ of a 88/56 battery scan the horizon with their binoculars while the crew of the fire control telemeter (in the background) get on with their job. (J. Mata)

Following page: an F/88 artillery observer. (Via the authors)

Previous page: A German Second Lieutenant poses on the running board of a light car. (Campesino, via R. Arias)

Above: The complete crew of a 88/56. (Authors)

Below: The position in a battery with the most visitors come the end of the day. As can be appreciated, luxuries were not in abundance during the war. (Via the authors)

Above: German soldiers enjoy their time off with a beer. (AGGC)

Below: The "Generalisimo" making a visit to the Headquarters of the Condor Legion. Behind him is General Volkmann, then commander of the unit. (Buch, via R. Arias)

Light 2 cm Flak 30

Heavy 8.8 cm Flak 18

Above left: Colonel Hermann Lichtemberger
Above right: Colonel Georg Neuffer
Below right: NCO of the F/88
Decorations: Cross of Spain, the Campaign Medal, and the War Cross 36-39

Light 3.7 cm Flak 18

Above: 88/56 German anti-aircraft gun Flak 18.

Below: Fire control telemeter 36 for anti-aircraft batteries of 88/56 mm.

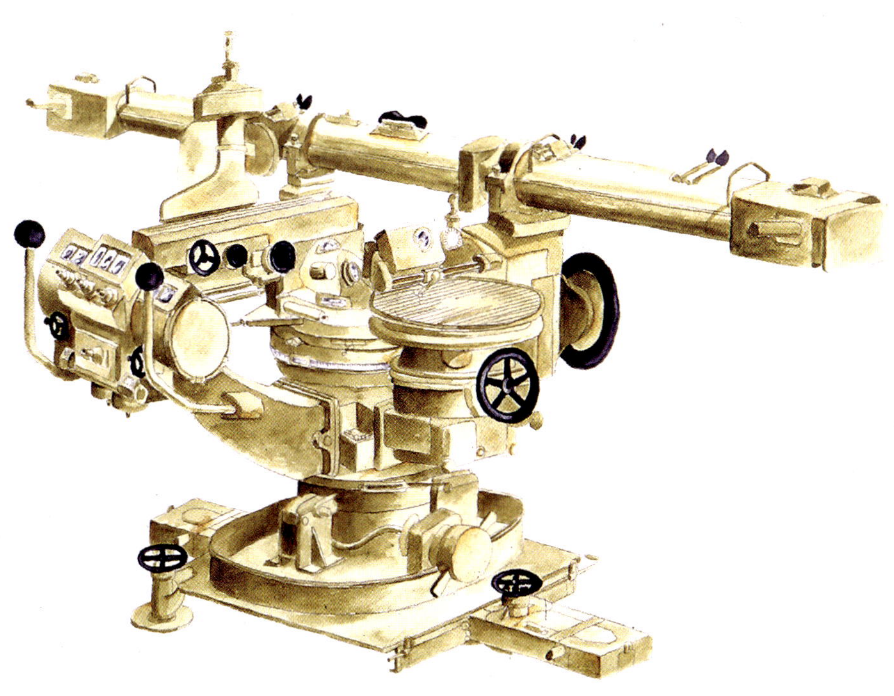

Cantabrian Sea
MAR CANTÁBRICO
OCÉANO ATLÁNTICO
Atlantic Ocean
La Coruña
Oviedo
Santander
Río Miño
Vigo
Pontevedra
León
Zamora
Río Duero
Salamanca
Segovia
Ávila
Brunete
PORTUGAL
Río Tajo
Toledo
Cáceres
Río Guadiana
Río Guadalquivir
Córdoba
Sevilla
Málaga
Cádiz
Ceuta
ISLAS CANARIAS
Canary Islands
Santa Cruz de Tenerife
Las Palmas de Gran Canaria

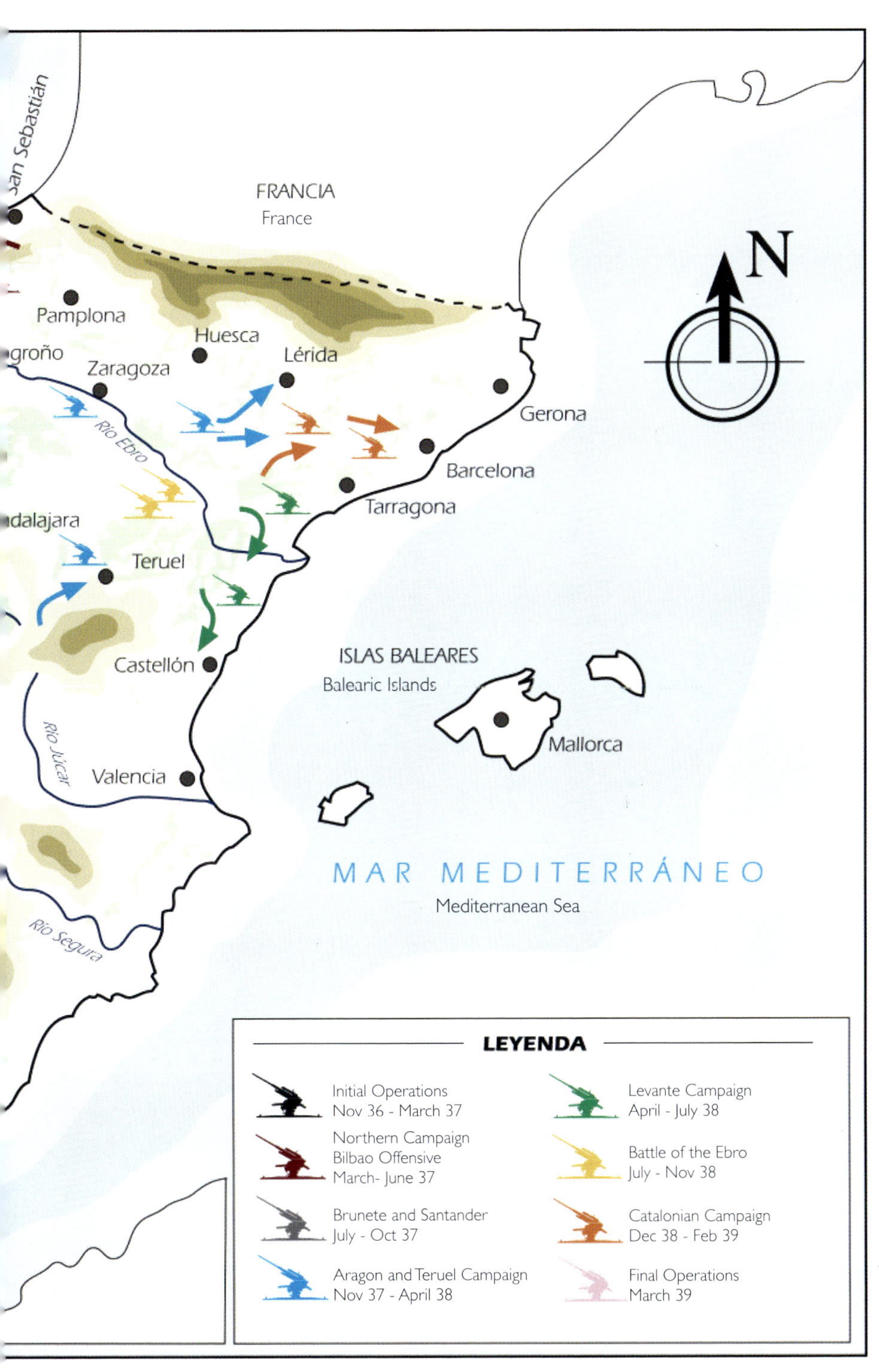

San Sebastián
FRANCIA
France
N
Pamplona
Huesca
groño
Zaragoza
Lérida
Río Ebro
Gerona
Barcelona
Tarragona
dalajara
Teruel
Castellón
ISLAS BALEARES
Balearic Islands
Mallorca
Río Júcar
Valencia
Río Segura
MAR MEDITERRÁNEO
Mediterranean Sea
LEYENDA
Initial Operations
Nov 36 - March 37
Northern Campaign
Bilbao Offensive
March- June 37
Brunete and Santander
July - Oct 37
Aragon and Teruel Campaign
Nov 37 - April 38
Levante Campaign
April - July 38
Battle of the Ebro
July - Nov 38
Catalonian Campaign
Dec 38 - Feb 39
Final Operations
March 39

Heavy 7.5 cm Flak 14

6

The Campaign in the Levante

From 20 April 1938 three heavy batteries and one light battery of the Condor Legion were attached to the Galicia Army Corps in order to take part in the campaign in the Levante. This Army Corps was composed of four divisions, the 4th, 55th, 83rd, and 84th. One heavy battery, half a light battery, and a troop of 37mm cannons were deployed around the German fighter aerodrome at La Cenia (the rest of the fighters were in Zaidin and the bombers in Zaragoza), and another heavy battery and half a light battery were deployed in Benicarlo, seat of the German Tactical Headquarters (General Volkman).

Several members of F/88 resting in one of the tents that served as sleeping quarters at a Nationalist aerodrome. (Campesino)

Junkers Ju 87 B Stuka. The Condor Legion was equipped with a unit of these magnificent dive bombers. (Calparsoro)

The attack commenced on the 23rd, with the Castille Army Corps (commanded by General Varela) advancing through the interior on a north-south axis, while the Galicia Army Corps advanced along the coast. A heavy battery took part in the fighting on the first day, and the following day there were two heavy batteries and one light battery engaged, including actions against occasional aerial targets. On the 25th three heavy and one light batteries were involved, with the German guns being used in the counter-battery and anti-tank roles, opposing four counter-attacks as well as attacks from the air. The next day (26th) the two heavy batteries were to be involved in the heavy fighting experienced by the 83rd Divison in the vicinity of the well entrenched area of Cuevas de Vinroma, even involving combat with enemy anti-aircraft batteries. The day also saw the Ju 87s, the famous Stukas, in action .

The heavy rain and the determined enemy defence imposed a lull in F/88's operations until 3 May, when the German gunners exchanged fire with enemy artillery. On the 4th three heavy batteries and one light battery were in action, especially around Albocacer, with 2.F/88 receiving counter-battery fire. The next day was almost identical, but this time with 3.F/88 coming under attack from counter-battery fire. On the 6th, in the hard fighting south of Vinroma and Alcala de Chivert, it was 1.F/88 that came under enemy artillery fire (curiously, these attacks by enemy artillery produced no casualties). The 5.F/88 contributed to the capture of a rail mounted enemy battery, forcing another entrained battery and two field batteries to halt their fire.

However, on the 7th, while one battery was engaged against ground targets, there was a bombing attack by Tupoliev SBs at La Cenia and Vinaroz, to which the anti-aircraft guns were unable to effectively reply due to the cloudy conditions. Therefore, in midafternoon the Flak batteries supporting the Galicia Army Corps were withdrawn in order to reinforce those defending La Cenia, Vinaroz, and Benicarlo.

Armoured train and Republican artillery captured by the Nationalist Army. (Queipo de Llano, via J. F.-Coppel)

The first night attack by a solitary plane on the aforementioned towns (a technique the Russians were to make use of during both the remainder of the Spanish Civil War and later during World War II) occurred on the 12th at Benicarlo. The same occurred on the 14th, and on both occasions (the aircraft involved was probably an RZ "Natacha") both the heavy and light Flak batteries were unable to prevent the attacks. Another biplane carried out a similar attack on the 15th, managing to surprise the defences and escape, fired at only by the 20mm cannons. None of these incursions produced serious damage or casualties, but command of the skies in low visibility was in question, with the enemy aircraft escaping observation by infiltrating from the sea.

The German report from the 17th reflects on the action of 3.F/88 against six Tupoliev SBs and 20 Polikarpov I-16s, as well as that of 1.F/88 against another nine I-16s. The

Previous: A Flak 18 88/56 in service with the Spanish units.

Above: German soldiers belonging to F/88 chat with Spanish comrades from the Civil Guard next to an 88/56 mm Flak 18, La Cenia, 1938. (Heribert G., via R. Arias)

report adds that one battery received a hit from a bomb that injured one man and destroyed two motorcycles and a car. Jesus Salas afirms that there were no sorties by the Katiuskas that day, but he does state that their were Republican attacks on the aerodromes at Caude and La Cenia on the 18th, so it is highly likely that these attacks were in fact on the 17th, although they are mentioned in a written document dated the following day.

The Condor Legion returned its efforts to providing support to the Galicia Army Corps on the 18th and 19th with three heavy batteries supporting the attack by the 4th Division with the shelling of ground targets, receiving enemy counter-battery fire in the process. On the night of the 18th, the light and heavy batteries defending the German Headquarters at Benicarlo repulsed one of the by now familiar lone air raids, doing so again on the 20th. The night visitor had now been given the nickname "Night Goose." On this and the following day a heavy battery destroyed targets on the ground at Cati and Nevera.

On the night of the 24th the "Night Goose" returned to lay her "eggs" in Benicarlo, and on this day F/88 was reorganised for combat; the new structure was as follows:

- Support and air defence for the Galicia Army Corps (Aranda): 2 heavy batteries and half a light battery.
- Air defence of the Condor Legion's command post in Benicarlo: 1 heavy battery and half a light battery.
- Defence of the La Cenia aerodrome: 1 heavy and ½ a light battery.
- Defence of the port of Vinaroz: 1 heavy and ½ a light battery.

The imposing Sd.Kfz 7 KM m11 tractors, towing the ubiquitous 88/56 Flak 18. (J. Mata)

The solitary night attacks from the sea were repeated again on the 27th. During the following three days support was considered for those formations under General Aranda, namely the 4th Navarra and the 55th Division, from the "Alonso Vega Group," in their attack on Ares del Maestre. Nine Tupoliev SB bombers and their escort fighters attacked La Cenia

From right to left: 2nd Lieutenant Bonader and Warrant Officers Herbert Beyer and Heinz Dreger, members of I.F/88. (Campesino)

on the 30th, although the attack was largely ineffectual due to the fact the bombs were dropped a kilometer from their target and the anti-aircraft batteries were not engaged.

May ended with a battery of 88s supporting the 55th Division, and June began in a similar vein. The Castille Army Corps ceased its attacks at this time.

June 2nd saw an attack by nine Katiuskas belonging to the 3rd Squadron 24th Group at La Cenia, with five planes being lost to fighters and anti-aircraft fire, and a gunner from the 5.F/88 wounded. According to Andres Garcia Lacalle , the observer-navigator of one of the attacking planes, Fernando Medina Martinez, informed him that:

"...two planes were lost from direct hits by the German anti-aircraft artillery and the other three had been hit by the anti-aircraft artillery and finished off by the Bf 109s...."

An observer from one of the Tupoliev SBs, Justo Martin Calderon, saw how a direct hit from the anti-aircraft guns tore off the wing of one plane, which immediately went down in flames (the pilot, Miguel Baile, had no time to bail out). Curiously, the German report did not claim these planes as having been destroyed by its anti-aircraft artillery, acrediting all the kills to the fighters, although this was common between fliers and gunners, especially when the former were responsible for making the report.

German Messerschmit Bf 109 fighter. The aircraft shown belonged to 2.J/88 Squadron; the plane in the background bears the unit's "Zilinder Hat" emblem. (Campesino)

Fernando Medina Martinez described his experience as a navigator in a Tupoliev SB piloted by Jose Miñana Juan when they were hit at 6,000 m by flak before being finished off by fighters. The Nationalist report accounts for two Katiuskas brought down by its 41st Battery, deployed in the vicinity of San Mateo, along with the capture of prisoners (in Tirig, 30 km to the southeast of La Cenia, on the return flight path of the bombers, which had approached from the sea). This unit fired forty-six shells that day, claiming two planes shot down. Salas also attributes two to the anti-aircraft batteries. In summary, two of the planes shot down can be credited to the German flak, as well as contributing to the loss of another three brought down either by German fighters or Spanish anti-aircraft gunners.

The following day 24 Tupoliev SBs, escorted by 39 fighters, attacked the Condor Legion's command post in San Cristobal. In this case there were no planes lost and very little damage reported. The two previously mentioned heavy batteries continued to support the 4th and 55th Divisions on this and the following day, with a battery being bombarded by enemy artillery, although with no losses in men or materiel.

A fine photograph of a Soviet Tupoliev SB, called the "Katiuska" in Spain, and a Martin Bomber. (C. Azaola)

A 88/56 mm anti-aircraft battery. The guns are mounted on their carriages ready for towing. (Wilhelmi, via the authors)

On the 5th the operations report of the Condor Legion succinctly affirmed that their 88mm batteries repulsed an attack on La Cenia and gave support to the 4th Division. There are references to another heavy battery that opened fire on six bombers. The next day brought more of the same, and personnel from 4.F/88 came under attack from enemy ground fire.

Before continuing it is necessary to mention the deaths by firing squad (the circumstances of their capture remain unknown) of gunner Walter Müller on the 2nd in Bechi and the NCO Wolfgang Bauer in Castellon on the 6th, both members of F/88.

On the 7th a German battery opened fire on seven light bombers and forty-five I-15 and I-16 fighters; another two heavy batteries supported the 4th Division, engaging tanks and armoured vehicles as well as Republican batteries. The next day both batteries of 88s fired on six Tupoliev SBs and nine I-16s that were attacking Nationalist troops. Meanwhile, a night bomber was engaged in the rearguard. The two F/88 batteries continued to support troops in the front line on the 9th.

The 10th was particularly difficult for F/88, with two heavy batteries and one light battery supporting the 4th Division, which was infiltrating west, and also protected the Division from air attacks in the area of Azdaneta, resulting in two Katiuskas shot down by 6.F/88 and an I-15 by 4.F/88. Over the next few days the German batteries continued to provide ground and anti-aircraft support, with 6.F/88 claiming a Tupoliev SB brought down on the 13th.

Previous page: a light 37mm Flak 18 in firing position on its carriage. (J. Negreira)

Above: a Soviet Polikarpov I-16 "Mosca" (also called a "Rata" by the Nationalist troops). (F. Ezquerro)

Conical tents used to acommodate the crews of F/88 defending Nationalist aerodromes. (Campesino)

The Nationalists took Castellon de la Plana on the 14th, and F/88 continued to give the same system of support as before, with the 6.F/88 claiming two Polikarpov I-15s in Villareal on the 15th and the 4.F/88 another I-15 on the 16th; all were later confirmed by the Nationalists.

It can be appreciated from this description that this was harder and less disorganised than the earlier Battle of Aragon, with the Republicans having successive well prepared lines of defence occupied by fresh troops (the famous "X-Y-Z" line), and using their reserve troops who had been told to hold at all costs in a much more rational way, supported by a commited Air Force. All this meant that the Fak had to be employed in depth, with risks not only from the air, but also on the ground.

In the interests of offering a simplified version of these events, the actions of F/88 during the months of July and August up to the beginning of the Battle of the Ebro have been summarized in the following table of events. In July a troop of 37mm guns were attached to the Liaison Detachment (General Garcia-Valiño) for the failed assault on the Sierra del Espadan.

Levante: The German Flak Battalion, June to July 1938

Day	Participation of F/88 in air-ground combat
17-JUN	30 Polikarpov I-15s engaged. Two batteries of 88s and half a light battery join the pursuit.
18-JUN	The 88s fire on enemy planes. During the night 3F/88 is attacked, suffering varied damage.
19-JUN	Night attacks on rearguard. Two heavy batteries harass the enemy.
20-JUN	Two batteries of 88s and two of 20mm engage enemy fighters and bombers.
21-JUN	One 88mm battery and half a 20mm battery fire on machine-guns and observation posts.

The 88mm gun proved itself to be a formidable weapon during the Spanish Civil War, capable of accurate fire against enemy aircraft as well as ground targets such as vehicles, moving tanks, bunkers, and machine-gun nests. (Campesino, via R. Arias)

23-JUN Bombardment of enemy field battery.
24-JUN Harrassing fire on enemy retreating by road.
25-JUN Three 88mm batteries engage enemy planes.
26-JUN Corporal Ehrhardt Horn of 2.F/88 killed and two officers wounded by enemy shell fire (a Major and a Second Lieutenant).
29-JUN At night, following an intense day of counter-battery and harassing fire, the 2nd battery is heavily bombarded by the enemy, but no casualties are reported.
30-JUN Three batteries of 88s engaged in fire support and counter-battery fire.
1-JUL Two 88mm batteries in action, firing against railroad infratructure and counter-battery fire (Nules). Enemy counter-battery fire, but no casualties reported.
2-JUL A battery of 88s provides ground support and counter-battery fire (Villavieja, Burriana, etc). Gunner Walter Müller executed in Bechi by a Republican firing squad.
3-JUL Five SBs, thirty I-15s, and six I-16s fired on, counter-battery fire, and deny road to enemy traffic.
4-JUL An SB engaged at the front. Two batteries of 88s provide ground support, counter-battery, and harassing fire.
5-JUL Two batteries of 88s attack targets in Burriana-Nules-Arena.
6-JUL Two batteries of 88s provide ground support (Monfocar-Chilches, Alfondeguilla, north of Val de Uxo). The NCO Wolfgang Bauer executed in Bechi by a Republican firing squad.
7-JUL A battery of 88s engages enemy aircraft. Fire support given to the attack by the 4th Navarra Division.
8-JUL Two 88mm batteries and one 20mm battery fire on six SBs and 15 Polikarpov I-15s. Two 88mm batteries harass enemy rearguard. Sergeant Erich Vandrey killed by enemy artillery fire.
10/ 15-JUL A light battery in the rearguard in action against a night bombing attack.
18-JUL Two men wounded by shrapnel.
23-JUL A heavy bomber fires on six "Boeing" bombers, bringing down one confirmed and another probable; Nationalist documents support the claim, also indicating it happened in the Artana sector. On the same day the Nationalist Army's 28th battery claimed an SB shot down in the Sierra de Onda. A Sergeant from 5.F/88 was wounded by incoming shrapnel.

7

F/88 and the Battle of the Ebro

The Battle of the Ebro began on the dawn of 25 July 1938, at the same time the Nationalist divisions were advancing on Valencia and the Sudetenland Crisis continued in Europe. The 50th Division of the Moroccan Army Corps (General Yagüe), which was occupying the area that came under attack, was over-run, and units from other fronts had to be rushed in to stop the gap. This resulted in the immediate paralysis of the Levante offensive, and the Republican advance was stopped at Gandesa on the 29th. The attackers had penetrated to a maxim depth of 25 km along a 35km front, taking an area covering some 600 km^2 between the rivers Matarraña and Canaleta, including mountainous zones (Sierras de Fatarella, Pandols and Caballs) that would become famous for the fighting that took place during their reconquest.

On the 29th those batteries of F/88 supporting the Galicia Army Corps left the Levante and deployed on the new front south of the main bridgehead (Gandesa, Pandols, Río Canaleta), in support of the 4th and 84th Divisions. On the same day a Sd.Kfz 7 tractor of 1.F/88 accidentally caught fire. Two days later the Group entered combat, with a light battery firing on 23 Polikarpov I-15s and a heavy battery against tanks and vehicles between Pinell and Cherta. On 3 August the 6.F/88 fired on thirteen Katiuskas and four I-16s, claiming a Katiuska shot down near Gandesa. A light battery fired on ten SB bombers but recorded no hits. The next day the 88s continued firing, targeting vehicles and light anti-aircraft batteries on the road from Pinell to Cherta, as well as six Katiuskas.

The enemy had been contained by the 5th and the front stabilized, and there then followed a battle of attrition that lasted until 29 October.

Following a period of inactivity, a heavy battery was again involved in providing fire support for the 4th Navarra Division against the Republican 11th Division on 9 August in the area around the Santa Magdalena Heights (Sierra de Pandols) and the Gandesa to Pinell road. At the end of that month, and following the total exhaustion of those two magnificent units, the Santa Magdalena hill continued to be in Republican hands. Casualties on both sides gave testimony to some of the hardest fighting of the war. A summary of the involvement of the Condor Legion's Flak Group that August reflects its support of the Moroccan Army Corps, involving the Group in both the anti-aircraft and ground support roles (fire support, counter-battery fire, harassing fire, destruction of river crossings, etc).

The Germans were able to put their equipment to the test in the Spanish Civil War, allowing them to modify, cancel, or improve many of the projects being developed by German private industry. (Campesino, via R. Arias)

F/88 in the Sierra de Pandols, August 1938

Day	Participation of F/88 in air-ground combat.
10-AUG	A heavy battery supports the ground attack; also bombards crossings over the Ebro at Benifallet
11-AUG	A heavy battery supports the infantry attacks.
12-AUG	A heavy battery supports the ground attack. The 88s engage five Katiuskas.
13-AUG	88s used against machine-gun nests. Heavy guns fire on 15 Polikarpov I-15s.
14-AUG	A heavy battery supports attack. Three batteries of 88s and one of 20mm cannons engage 35 Polikarpov I-15/16 fighters, 30 I-16, and 6 MB/Katiuskas.
15-AUG	Three heavy batteries fire on five "Ratas" and four SBs. The 35th Division is relieved that night.
16-AUG	A battery of 88s bombards an enemy field battery.
17-AUG	Three batteries of 88s and half a 20mm battery fire bring down 35 I-15s and I-16s. One heavy battery engages enemy artillery and another supports the infantry attacks.
18-AUG	A battery of 88s engages an enemy battery.
19-AUG	Three 88mm batteries support the land attack. The 4th Navarra is relieved by the 84th Division. A heavy battery fires on fifteen I-16s.
19 to 27-AUG	Offensive against Cuatro Caminos/Fatarella, Gaeta Heights, north of Gandesa.
20-AUG	Three heavy batteries support the ground attack.
21-AUG	Three batteries of 88s and half a 20mm battery fire on nine SBs and twenty I-16s.
22-AUG	Three heavy batteries engage ground targets. Two batteries of 88s and half a 20mm battery open fire on three occasions against eight Katiuskas and 30 I-15/16s. Gaeta Heights occupied.
23-AUG	Three batteries of 88s and half a 20mm battery fire on seven Tupoliev SBs and 30 Polikarpov I-15s and I-16s. A Tupoliev is shot down.
24-AUG	Ground support from three 88mm batteries. Three batteries of 88s and half a 20mm battery fire on seven Katiuskas, nine I-16s, and a score of I-15s. A I-16 is shot down (Gascon, Badia, Calatayud?).
25-AUG	Three heavy and one light battery provide ground support. A German Warrant Officer is wounded by enemy fire.
26-AUG	Six Tupoliev SBs and 25 I-15/16s engaged. Two 88mm batteries involved in counter-battery fire.
27-AUG	Ground targets attacked by four batteries. One supports the 84th Division and another engaged in counter-battery fire.

28-AUG	Two 88mm batteries fire on enemy ground positions.
	One 88mm and half a 20mm battery engage thirty-five fighter planes.
29-AUG	One heavy and half a light battery attack enemy positions and artillery. A German gunner is wounded by shrapnel.
30-AUG	Two troops of light artillery come under enemy counter-battery fire: A truck damaged.

With the offensive on the Sierra Fatarella having run out of steam the Nationalist command returned to the offensive, attacking the enemy centre (Sierra de la Vall-Venta de Camposines -Los Gironeses) following the reorganisation of the units comprising the Army Corps. The Moroccan Army Corps, commanded by General Juan Yagüe, was made up of the 50th, 82nd, 152nd, and 4th Divisions, deployed from north to south as far as Coll de Grau.

The Maestrazgo Army Corps (a newly created formation under the command of General Garcia Valiño) included the 1st Navarra Division (General Mizzian) and the 84th and 74th Divisions.

Right and above: Gunners of F/88 were able to test the latest designs in military technology under combat conditions. The 88/56mm anti-aircraft gun was the reference in anti-aircraft defence at the time. (Campesino, via R. Arias)

The 13th Division was deployed between the 4th Division and the 1st Navarra as the Army of the North's reserve.

On 1 September General Davila issued the operation orders for the attack, which was to be preceded by a four-hour artillery barrage (two of them for fire adjustment) from 76 batteries, among which were those belonging to F/88, three heavy 88mm batteries, and one light 20 mm battery.

The German Flak Group changed its positions in order to support the Moroccan Army Corps, and from the 1 September a heavy battery was engaged in counter-battery fire northeast of Corbera, in the area earmarked for the attack of the 4th Navarra Division, while two light troops were bombarded by enemy artillery. On the following day a battery of 88s opened fire on two occasions, once against six Katiuskas, and another time against twenty I-16s and twelve I-15s, while the 4.F/88 was bombed.

A Krupp L-2 H 43 Protze tractor towing a light 20mm Flak 30 at an airfield. (Via authors)

On the 3rd, the three heavy and one light German batteries joined in the initial barrage. Activities on the 4th were mainly concerned with anti-aircraft fire, with a heavy battery engaging six Tupoliev SBs, all four batteries firing at 30 fighters, and two batteries of 88s and half a 20mm battery engaging 25 Polikarpov I-15s. The 4th Division took hills 385 and 408 the next day supported by one of F/88's heavy batteries, while two others attacked troop concentrations and traffic.

Meanwhile, other heavy guns from F/88 repulsed attacks from a dozen I-16s on the first day and by another thirty fighters the next. On the 6th the German batteries were occupied with counter-battery fire and targeting armoured vehicles, as well as being in action against several concentrations of fighters. During the days that followed the 1st Division completed the occupation of the Sierra de la Vall, while the German gunners fired at anything that moved, both on the ground and in the air, destroying trenches, strong points, and Republican anti-aircraft batteries. On the 8th the Nationalist offensive found itself in crisis.

Over the following days until the 13th the enemy counterattacked, preceded by air attacks and artillery bombardments that at times lasted more than two hours, which brought about the end of this phase of the battle. The German gunners continued to be active in both anti-aircraft defence and ground support missions, claiming the previously mentioned Polikarpov I-15s on the 13th.

Cleaning the tube on a 88/56mm gun destined for the defence of an aerodrome. The 250 Kg bombs denote another aspect of the air war. (Via Marin)

A gunner of F/88 holding a 88mm shell poses in front of his anti-aircraft gun. (Campesino, via R. Arias)

The Flak did not see any combat between the 14th and 17th due in part to bad weather. It is also necessary to mention that during the Sudetenland crisis, the French were pondering the idea of invading Spanish Morocco and the Pyrenees, making it necessary to send troops to these areas and fortify them.

A new operations order was issued by the Army of the North on the 18th that assigned the 13th Division the task of taking the Venta de Camposines. The Division was to be flanked on the right by the 1st Division and on the left by the 4th. This period of operations was to last until 14 October, with the Nationalists achieving only small advances. Furthermore, this period coincided with the rumour that London and Paris planned to intern the personnel of the Condor Legion should war break out; there was also the arrival of heavy rains at the end of September, the proposal by Doctor Negrin, head of the Republican government, to repatriate the International Brigades (they fought at the front for the last time on 23 September), and Franco's proclamation on the 27th that Spain would remain neutral should there be war in Europe. The crisis in Europe reached its zenith on 29 September with the signing of the Munich Agreement, and on 1 October Germany annexed the Sudetenland. Franco took the decision to discharge 10,000 Italian troops, combining those that remained in the "Littorio" Division and the Hispano-Italian "Flechas," although he retained the artillery and tanks of the CTV.

F/88 was to be equipped with a total of five batteries of 88s, with four guns allocated to each battery. (Wilhelmi, via authors)

Once the crisis in Europe had abated Germany gave the green light to renew the shipment of arms to Spain, as well as relieving personnel of the Condor Legion. The new commander was General von Richthofen (Sperrle's former Chief of Staff), aided by Colonel Seidemann as Chief of Staff.

The activities of the German anti-aircraft gunners during this period, including the eve of the attack on the Sierra de Caballs, are summarized in the following table of events.

F/88 and the second attempt at taking Venta de Camposines, September to October 1938

Day	**Participation of F/88 in air-ground combat.**
18-SEPT	Three heavy batteries support the attack. 24 I-15s and 11 I-16s engaged, and nine I-15s and 26 I-16s on another front.
19-SEPT	Two heavy batteries engage two enemy batteries. A battery of 88s fires on 26 I-15s and 19 I-16s.
20-SEPT	Three 88mm batteries fire on 24 I-15s and 18 I-16s; they also fire in support of ground forces. Significant parenthesis coinciding with the Munich crisis.

The new commander of the Condor Legion, General von Richthofen, leaving with an officer of S/88 and with the still Chief of Staff, Plocher. (Campesino, via authors)

2-OCT	Two 88mm batteries support the 1st Division occupying the Sierra de la Vall. A battery of 88s and a light battery fire on three I-15s (one confirmed shot down another probable).
5-OCT	Counter-battery and ground support by three heavy batteries.
7-OCT	A battery of 88s claimed one of six Polikarpov I-16s shot down.
8-OCT	Two heavy batteries and one light battery fire on six Katiuskas escorted by fighters, with the light battery engaging up to forty fighters. The batteries defending the base at La Cenia fire at six Dewoitines. Two 88mm batteries engaged in counter-battery fire at the front.
10-OCT	A heavy battery harasses road at Venta de Camposines. Two more engaged in counter-battery fire. One 88mm and one 20mm battery fire on five Katiuskas.
11-OCT	Counter-battery fire by one battery of 88s.
14-OCT	Five Katiuskas and several fighters engaged. One Soviet bomber claimed as destroyed and another probable.
18-OCT	Two 88mm batteries engaged in harassing and counter battery fire and destruction of ground targets.
20-OCT	A battery of 88s engaged in counter-battery fire in the Pinell sector, denoting a change in deployment in the south in the area of the Sierra de Caballs; another battery fires on an I-15. Much rainfall.

The operations against the Sierra de Caballs were begun on 30 October from the positions in the Sierra de la Vall. More than 340 guns belonging to 95 batteries, including the 4 from F/88, took part in the opening bombardment, which began at 0700 hours (with three hours of adjustment and one of execution). There then followed bombing attacks by more than one hundred aircraft. The main attack was to be made by the 1st Division of the Maestrazgo Army Corps, while the formation's other three Divisions (the 84th, 74th, and 82nd) cooperated in the action. The main burden of defence fell on the Republican V Army Corps' 43rd Division (the same Division from the Bielsa pocket that withdrew towards France). The Sierra had been as good as taken by mid-day, and by the 2 November the whole Pandols massif had been conquered, on the 3rd the village of Pinell, on the 7th Mora de Ebro, and by the 16th the Battle of the Ebro was over.

Meanwhile, on 31 October 8.F/88 reported the probable destruction of a Polikarpov I-15 in the vicinity of Salvatierra. On the night of 5 November two heavy batteries opened fire on six Katiuskas while a third battery engaged another, resulting in another probable enemy aircraft destroyed near Camposines-Mora. Five days later a light battery fired at a group of 20 fighters, claiming two probable I-15s brought down in the Nules sector, reported later as confirmed in the Air Force Staff's monthly bulltin.

Following the conclusion of the Ebro battle there was a lull of almost a month before the start of the Catalonian campaign, which began at the end of December. During this time the most important action undertaken by the Condor Legion was probably the defence of the La Cenia aerodrome on 16 November, in which the Legion's official report states that an attack by 10 Katiuskas was repulsed, with one shot down by the Flak and another by a Messerschmitt Bf 109 piloted by Herbert Schob, with the fighter later pursuing two other bombers. It is worth going into more detail on this action, as there are details that require closer attention.

Farewell celebration for the commander of I.F/88, Captain Haizmann, held in the emplacements of the Nationalist Army's 40th Anti-Aircraft Battery in October 1938. (C. Franco)

Nine Katiuska bombers (one from the HQ of 24th Group piloted by Lt Colonel Leocadio Mendiola, three from 2nd Squadron under Captain Francisco Gomez, and five from the 4th commanded by Captain Jaime Mata) took off at dawn heading for Mallorca. After turning towards Castellon they flew over Vinaroz at 7,500 m, diving towards La Cenia in order to gain speed and drop bombs from an altitude of 3,500 m, completing their bombing run, according to Mendiola, in open order in a north to south direction, which meant they took all the necessary precautions to avoid detection from the Nationalist observation posts.

The first group of three planes was that of Mata and Mendiola, the second was Gomez, and the third Celra. The shells from the 88/56 Flak 18s hit the second group, hitting Gomez-Ballester's plane full on, forcing it straight into a dive, and then the left wing engine of Gomez-Ballester's right-hand wingman Ricondo, whose plane lost height and was finished off by the Bf109.

Emplacement of a 20mm Flak 30 from one of the light batteries of *Flakabteilung* 88. The Germans were aware of the validity of the concept of this weapon, but at the same time they detected failings and complications present in the equipment. (Campesino, via R. Arias)

However, what is really curious for us today are Mendiola's comments on the incident in the magazine "Aeroplano":

"...they had all their air space covered by modern anti-aircraft artillery, (...) using the tactic of creating cones of fire at three different altitudes at the same point, all controlled by a primitive but effective radar (...) we were met by a terrible wall of anti-aircraft fire (...) later I found out they had followed us on their radar from take off to La Cenia."

Mendiola was possibly quoting Soviet sources, and it is also known that the *Luftwaffe* had a mobile air surveillance radar system operational in the Autumn of 1939 known as Freya, which had been officially presented to Hitler in June 1938 after having undergone trials in the large-scale German manoeuvres in the Autumn of 1937 in Swinemünde, together with another similar system designated Seetakt.

Germany had spurred on its research into this field in 1933, the year Hitler came to power, mainly by the *Kriegsmarine*, with Rudolph Kunhold (from 1929) developing in

A heavy 88/56 mm Flak 18. (Wilhelmi, via authors)

1934 what would be the first radar in the world, capable of detecting a ship at a range of 12 km. In 1935 the Telefunken engineer Wilhelm Runge recorded echoes from a plane at 5 km with another model, the Dezimeter-Telegraphie (DeTe), or radio direction finder. A little later, the *Kriegsmarine* installed the Seetakt radar on the "Bismarck" and the "Admiral Graf Spee" to test its use in zeroing naval gunfire, working at a frequency of 375 MHz (80cm wavelength) and with a range of 9 miles. This equipment was used by German ships operating along the Spanish coast during their patrols to enforce the "No Intervention" policy agreed to by the League of Nations.

Whatever the case, and in the absence of specific documents proving its employment in Spain, it is not unreasonable to suggest that given the fact the Germans employed some of their latest experimental materiel in Spain, such as the Junkers Ju 87 Stuka, the 88/56 Flak 18, the 105/52 and 105/26 field guns, sound-locating equipment, and even their Enigma machine, to give just a few examples, that they also used the occasion to secretly test their radar.

A Henschel Diesel truck belonging to F/88. (J. Mata)

8

The Catalonian Campaign and the End of the Spanish Civil War

On 23 December the Urgel and Maestrazgo Army Corps broke through the enemy lines at the bridgeheads at La Baronía and Tremp, initiating the start of the Catalonia campaign. The Condor Legion was given the order to support the Navarra Army Corps (General Solchaga, 4th, 5th, and 12th Divisions) and the Moroccan Army Corps (General Yagüe, 13th, 50th, and 105th Divisions) from its bases on the Sanjurjo, La Cenia, and Zaidin aerodromes, and its Command Post in Almacellas, as well as carrying out night bombing missions. In January it concentrated on supporting the Navarra Army Corps.

Years later, commenting on that campaign, the then General Martinez de Campos wrote:

"...The Condor Legion dismissed the notion that the 88mm anti-aircraft guns were not appropriate for employment on the ground; they did so with great accuracy, hitting moving cars that carelessly strayed into their field of fire...."

At this point in the war the Republican Air Force had been beaten, providing fewer opportunities for the Nationalist anti-aircraft artillery to fire at them. Throughout the Catalonian campaign, 93% of the shells fired by the anti-aircraft artillery were at ground targets. Concerning this fact Hidalgo Salazar writes of the brilliant cooperation between

Above: a light Krupp H-2 tractor belonging to the Condor Legion. (Campesino, via R. Arias)

Next page: A photograph of the Condor Legion's Anti Aircraft Group published in a Spanish booklet on the occasion of their redeployment to Germany. (Authors)

the German anti-aircraft artillery and the Navarra Army Corps as they advanced across Catalonia:

"Without any real need for anti-aircraft artillery, the German gunners were employed as close support artillery, noting in particular their skill in destroying several Russian tanks."

According to the authors Riess and Ring, from day one of the offensive the troops of 20mm and 37mm guns distinguished themselves against ground targets, playing a decisive part in the attack.

A very interesting example is an account written by a German NCO commanding a 88/56, commenting on the fighting at the Balaguer bridgehead on 2 and 3 January 1939:

"...Three powerful Nationalist columns tried to break through the Red front at the Tremp, Balaguer, and Seros bridgeheads. After driving slowly along roads filled with men and materiel our anti-aircraft battery arrived at the place where we were to deploy, Hill 343, east of Cubells. From early morning our own 88s, as well as field pieces of varying calibre, fired unceasingly at the heights which, fortified with cement (sic) drums, were soon to form the objective of our infantry.

The Reds, who were feeling safe behind the imposing blocks, fought viciously, and despite the heavy barrage from the supporting artillery, they were able to halt the advancing Nationalist troops. The situation was becoming increasingly difficult. The Red aircraft had appeared, and from our position we could hear the bombs exploding in the front lines. The men began handling the guns, sweating as they moved the levers as mechanically as if they were part of the same machine. The tubes were elevated and the direction changed, following the flight path of the bombers. An adjustment, then another, movements multiplying in order to synchronize the pieces with the new readings. Then, one of the planes falls wrapped in flames. In the sky our fighters appear and the enemy aircraft flee. All this had lasted no more than a few minutes, but in the meantime, without our support the infantry had to abandon the attack. Valuable seconds pass, and an order is received and

the gun aimers bring the angle of the tubes down at an incredible speed. The crew do not rest, with one shell after another fired. The Reds respond like men possessed. The resistance is ferocious. They repulsed the attack a second time. New adjustments. It is difficult to put a shell in the loophole of one of these concrete drums (sic) from which the machine guns spray bullets without respite. Firing, continually firing. The guns and crews seem to breath as one to the rhythm of the guns' recoil. Then, the observer shouted target! And one, two, five times in unison, the battery's shells crashed into the fortification's loopholes, and from above came the Nationalist troops. Watching them advance and plant the Spanish flag on the first of the blockhouses was like a symbol of the promise that the liberation of Catalonia was close at hand."

On 8 January 1939 the Flak brought down a I-15 in the southern sector of the Catalonian front, and on the 12th a column comprising a Republican battery was dismantled and captured on the road between Montblanch and Valls. The Flak 18s of F/88 fired at them with high explosive shells with timer fuses, causing the crews to flee and abandon their guns and vehicles to the Nationalist troops. On the 11th and 12th *Gefreiter* Herbert-Joachim Knohr and *Obergefreiters* Erwin Plewe and Georg Sordon died from wounds received during enemy counter-battery fire in Esplugas de Francolí (Tarragona).

On the 15th of that same month, the 5th Navarra Division carried out another of its lightning advances, covering the distance between Valls and Tarragona aboard trucks. The tanks and anti-aircraft batteries supporting them were ordered to advance ahead without stopping. The anti-aircraft guns opened fire on the retreating Republican vehicles, creating such confusion that the gunners captured some machine-guns trying to oppose them. The 88s, unable to keep up with such a rapid pace, which demanded constant changes

German gunner assigned to a Spanish Nationalist battery taking up position. (Authors)

A heavy Sd.Kf 7 tractor loaded on a railroad wagon, ready for transport. The war in Spain was about to end. (J. Mata)

in position, ceded their place at the head of the advance to the lighter guns commanded by Lieutenant Deventer. A last and desparate enemy counter-attack was repulsed by the light battery, forcing them to leave their positions. The result was that the tanks, infantry, and artillery entered the Catalonian capital so unexpectedly that their fire in the rearguard forced the Republicans confronting the 105th Division approaching Tarragona from the south to withdraw. Lieutenant Deventer's battery then managed to hit some targets that until then had never been attacked by the anti-aircraft gunners: three ships attempting to leave the port, that were almost certainly hit by the 37/57s.

Solchaga's decision, taken with the advice and support of von Richthofen, to advance from Valls to Tarragona along two narrow roads was one of the most risky taken during the war. That same day (the 15th) Yagüe also made a prodigious motorized advance of 50 km between Tortosa and Tarragona. Again, the "Lightning War" was being premiered in Spain.

Activity the next day was restricted to three batteries cooperating with the 5th Division, attacking enemy positions northeast of Rodoñia-Albiñana and Vendrell, as well as on troop concentrations, batteries, and tanks near Visbal del Penedes. One of these batteries also engaged some 15 fighters of varying types.

F/88's activities on 22 January are also recorded, again operating in support of the 5th Navarra. Three heavy batteries fired on vehicles near Mediona, on enemy troops and tanks near Villafranca del Penedes, and against enemy artillery, causing a magazine to explode. Two batteries of 88s also fired on two Katiuskas.

On the 23rd the Flak shot down an I-15. That same day Barcelona was evacuated. It is known that the next day a heavy battery opened fire on enemy fighters, while another engaged an enemy battery and put down harassing fire to deny the enemy the use of the roads. Meanwhile, another heavy battery fired on the port of Barcelona and the radio station

War's end. Nationalist troops parade in the Plaza de Catalonia (Barcelona, 27-I-1939). (M.E. Yagüe Foundation)

at Montjuich, which responded by raising the white flag. Finally, a fourth was in action with the Navarra Army Corps, breaking up pockets of enemy resistance and harassing the retreating Republican forces, as well as on the high ground south of Rodoñia. On the 25th F/88 engaged enemy fighters and attacked enemy positions and vehicles. Barcelona fell the following day, which only left the German gunners the task of defending the city and its aerodromes.

The story of the German anti-aircraft group in Spain ends with the rather strange death of Corporal Heinrich Husemann on 4 February in Roda, who was killed by small arms fire; the circumstances continue to remain unknown.

The operations order for the breakthrough on the Central Front, dated 7 March 1939, detailed that the four batteries of F/88 would provide support. According to Riess and Ring, on 17 March F/88 was in action on the outskirts of Aranjuez against an enemy battery. During this last period of operations, on the 24th of that same month the 1st, 2nd, and 3rd

Organization of F/88, March 1939

Commander:	*Oberst* Erich Kretschmann.
1. F/88 (88/56 Flak 18s):	*Hauptmann* Bundt.
2. F/88 (88/56 Flak 18s):	*Hauptmann* Reuter.
3. F/88 (88/56 Flak 18s):	*Hauptmann* Von Jablonski.
4. F/88 (20/65 Flak 30s):	*Hauptmann* Vogel.
5. F/88 (20/65 Flak 30s):	*Oberleutnant* Wehla.
(There were some 37mm Flak 18s assigned to the light batteries).	
6. F/88 (Searchlights):	*Oberleutnant* Hübener.
7. F/88 (Munitions):	*Leutnant* Jacobi.
8. F/88 (88/56 Flak 18):	*Leutnant* Hacker.
9. F/88 (88/56 Flak 18):	?

Total: 20 88mm Flak 18s, 38 20mm Flak 30s and 6 37mm Flak 18s.

Batteries of F/88 were positioned to the east of Ermita de Bastilla, tasked with protecting the Navarra Army Corps and ending the war together with the same Navarra Brigades with whom they had begun.

These events combine to show the versatility of these anti-aircraft guns. On the ground they were employed in the opening bombardments for the Nationalist offensives (at least in Bilbao and the Ebro), and they accompanied the ground forces in defence, attack, and exploitation, as well as engaging in counter-battery fire and direct fire (especially in the anti-tank role), even against shipping. According to Martinez de Campos, the batteries' all-terrain half-tracks allowed them, in the fighting in Asturias and Santander, to bring the heavy 88s into action to support the infantry before the field artillery arrived, stuck as it was on the roads cut by the retreating enemy. Martinez de Campos goes on to describe the destructive effects of the guns in his book *The Employment of Artillery* (1941), stating:

"...the combined fire of the Legion's field artillery never produced the kind of effect that one would have expected from it, whereas the individual shells fired from the Flak guns

Members of the German Flak Group wearing the *Luftwaffe* uniform; they are heading for the concentration area for the Condor Legion parade in Berlin. (Campesino)

Veterans of I.F/88 in Doberitz preparing for the Berlin victory parade held in June 1939. (Campesino)

were able to resolve situations in which there was no point in calling for the concentrated fire of several Field Artillery Groups (...) the anti-aircraft artillery achieved good results against certain kinds of ground and maritime targets...."

As a result of its experience in Spain, the *Luftwaffe* modified its classic diamond-shaped deployment of four battery Groups of 88s, substituting this for a formation of three in line, with the possibility of giving direct fire at ground targets, while the fourth battery was held back in a reserve position, providing anti-aircraft defence for the unit being supported. This also resulted in batteries being formed that were dedicated to ground combat.

Captain Lorenzo (in the magazine "Ejercito" nº47, 1943) states that the batteries of 88/56 Flak 18s, along with the Model 36 fire control system, firing by day, managed an average of 399 shells fired for each plane brought down during the war. This number increased to 561 for all the batteries taken together, whether completely equipped or not and including night actions. Of course, it must be remembered that he was reffering to Flak 18s in Nationalist hands, but we can easily apply these figures to the German guns, which provides an approximate idea of the rate of anti-aircraft fire. What is missing is the final total for aircraft shot down; various German publications give figures of between 59-61 enemy aircraft destroyed, giving an average of 60 (the authors have documented 41).

F/88 left a total of 33 dead in Spain, killed in action or in accidents. The unit took part in the great parade in Barajas and the victory parade held along the paseo de La Castellana in Madrid. After bidding farewell to the city of Leon, with which the unit had formed a close association, they embarked in Vigo on 24 May on the *Wilhem Gustloff*, leaving the majority of their equipment in Spain. Once back in Germany they paraded in Berlin before the National Socialist heirarchy.

J. FISCHER

Appendix I: The Commanders of *Flak Abteilung* F/88

GEORG NEUFFER (1895-1977)

Georg Neuffer was born in Steinbach, Oberpfalz, on 18 April 1895. He joined the *3.Feld-Artillerie-Regiment* Prinz Leopold as a *Fahnenjunker* in August 1914, undergoing training in Jüterbog in October and November that same year. From March 1915 he served as a *Batterieoffizier* in the *3.Bayerischen Feld-Artillerie-Regiment* until October the following year, when he transferred to the *22.Bayerischen Feld-Artillerie-Regiment*. After attending the *Flak-Schule* in Blankenberge and the *Flak-Kraftfahr-Schule* Valenciennes, he took command of *Heeresflak-Batterie* 54 (a Bavarian unit) in the middle of 1917. Later, he commanded *Flak-Batterie* 146 (also from Bavaria), and at the end of 1918 he was commanding a battery in the *3.Bayerischen Feld-Artillerie-Regiment*.

Between March 1919 and July 1920 he was *Batterieoffizier* in the *Volkswehr-Batterie* von Speck, being admitted to the Bavarian *Offizierschule* where he spent three months.

During the nineteen-twenties he found himself in several different artillery and infantry units (he attended the course at the *Infanterie-Schule* in Döberitz in 1924 and in the period 1926-1927 in the *Infanterie-Schule* in Ohrdruf) at the rank of *Oberleutnant* and *Hauptmann*. From October 1934 he spent a year as an instructor at the *Flak-Artillerie-Schule*. In April 1935 he was transferred to the *Luftwaffe*, havimg by then been promoted to the rank of Major.

In April 1937 he was appointed to command the *Flak-Abteilung* (mot) 88 in Spain, where he remained until 30 September the following year.

On his return to Germany he occupied a succession of different posts: commander of *I.Abteilung* of Flak-Regiment 64, of *Festungs-Flak-Abteilung* 34, and Flak-Regiment 14. He was promoted to *Oberstleutnant* in January 1938.

From April 1939 until just after the outbreak of World War II, he was posted to the *Reichsluftfahrtministerium* (RLM) in Berlin and in the *Stab General der Flakartillerie* under the command of the Air Force.

In October 1939 he was promoted to *Oberst*, becoming Chief of Staff of the *II.Flakkorps*, a large formation recently created in Frankfurt that would take part in the French campaign between May to June 1940.

In April 1941 he was appointed Chief of Staff of *Luftgau-Kommando IV*. In December he was promoted to *Generalmajor* and became Chief of Staff of *Feld-Luftgau-Kommando Moskau*, with its Headquarters in Minsk.

In April 1942 he was given command of 5.Flak-Division, deployed in Darmstadt, and in the following November that of 20.Flak-Division, formed in Leipzig and transferred to Tunisia to control all the units in that area of North Africa. In May 1943 Neuffer surrendered his unit to the British at the close of the North Africa campaign.

For the outstanding actions of the units under his command in the last phase of the fighting against enemy troops in Tunisia he was awarded the Knight's Cross on 1 August 1943, having been promoted to *Generalleutnant* one month before.

He remained a prisoner in British hands until June 1947. He died in the German town of Soest 11 May 1977.

HERMANN LICHTENBERGER (1892-1959)

Hermann Lichtenberger was born on 20 August 1892 in Germersheim. He joined the Army in October 1911. During the Great War he served first as an NCO, and later as an officer in Bavarian heavy artillery units. When the war finished he continued in the *Reichwehr*.

Although the Versailles treaty prohibited Germany from having anti-aircraft artillery, German officers trained in secret in managing the pieces and the tactics of air defence. Each artillery regiment had officers trained in these techniques, although the 7th Artillery Regiment (a Bavarian unit) had more than any other.

In 1935 the III Reich abandoned the Versailles Treaty and officially created the *Luftwaffe*, into which the Flak was integrated. Lichtenberger, at that time a Major, was given command of the first Group, Flak-Regiment 8 (I./Flak-Rgt. 8), which had been formed on 1 April 1935 in Fürth from the former *Flak-Abteilung* 7 (*Heer*). This unit came under the control of *Luftkreis* V.

While at the same rank, from April 1936 he commanded the first Group of Flak-Regiment 18 (I./Flak-Rgt. 18) (gem. mot.) based in Mannheim and forming part of *Luftkreis* IV.

At the beginning of August 1937 he was sent to Spain to command *Flak-Abteilung* 88 (mot.). While in Spain he sent numerous reports to Berlin on the employment of the anti-aircraft guns against ground targets (especially regarding the 88s), which were to prove extremely important later during the course of the Second World War. He was one of the sixty members of the Condor Legion to receive the Spanish Individual Military Medal. He returned to Germany in April 1938, where he returned to the command of I./Flak.Rgt. 18.

Oberst Lichtenberger commanded Flak-Regiment 5, which had been formed in Munich on 15 November 1938 and came under control of *Luftgaukommando* VII from November 1939 until June 1940.

At this same time Lichtenberger was given command of Flak-Regiment 104 (mot.), part of *I.Flakkorps*, and sent to Paris for Operation Sealion, the invasion of Britain. When the operation was abandoned in September the *Flakkorps*, along with Flak-Regiment 104, was transferred to Berlin to provide air defence of the capital.

In June 1941 *Oberst* Lichtenberger's unit was sent to Warsaw for the beginning of Operation Barbarossa. Once the attack was underway the unit was assigned to *Panzergruppe* 2, advancing in the center of the attack. For his outstanding conduct in command of the Regiment, *Oberst* Lichtenberger received the Knight's Cross on 12 November 1941 (bringing the number of recipients to 369).

A few days later he changed posts and passed to the *Führerreserve des Reichsministers der Luftfahrt und Oberbefehlshaber der Luftwaffe*, serving with *Luftgau* VII. He was to stay there until the middle of March 1942, when he was appointed inspector of the *Flakartillerie* in *Luftgau* Holland.

On 29 September he spent almost two months in charge of Flak-Brigade IV, deployed around Munich under *Luftgau-Kommando* VII. On 26 November he left his command (according to some sources due to illness) and returned to the *Führerreserve* until 21 September 1943, and was officialy removed from the active service list on 31 October 1943, at the age of 51.

He was by then a *Generalmajor*, having been promoted on 1 February 1943.

He died in the same town in which he had been born, Gemersheim, on 15 January 1959.

ERICH KRESSMANN (1891–1945)

Erich Kressmann was born on 1 September 1891 in Hofgeismar. He joined the *Heer* as a *Fahnenjunker* in October 1911. He served as an officer in the 20th Dragoons Regiment during the First World War and later on the staffs of several different units.

At the end of the war—by then a *Hauptmann*—he transferred to the Police, occupying several posts until October 1935, when he joined the *Luftwaffe* as a Major and *Abteilung-Kommandeur* in Flak-Regiment 4, a unit under *Luftkreis* IV .

On 1 March 1938, with the rank of *Oberstleutnant*, he took command of the Condor Legion's *Flak-Abteilung* (mot) 88 in Spain, relieving *Oberstleutnant* Lichtenberger, and staying in Spain until 30 September that year. On his return to Germany he was given command of Flak-Regiment 44, which formed part of *Luftverteidigungskommando* 4, deployed in Essen-Krag.

At the end of February 1940 he changed posts (he had been an *Oberst* since March 1939), this time taking command of Flak-Regiment 6, at that time stationed in Hamburg.

At the beginning of June he received command of Flak-Brigade II, integrated together with Flak-Brigade I in the *I.Flakkorps* (commanded by *Generaloberst* Hubert Weise). Kressmann's unit was formed in France on 1 June, and following the French surrender was deployed in the air defence role around Paris.

In accordance with the planned *Unternehmen Seelöwe* (Operation "Sealion," the invasion of Britain), he was charged with the air defence of the French channel coast and ports during the embarkation, including the transport ships,troops, and materiel.

When the project was abandoned, the unit was transferred to Berlin in October 1940 in order to defend the capital from air attack. In August 1941 Kressmann was appointed *Höherer Kommandeur der Luftgau-Flakartillerie-Schulen*. It was while in this post that he was promoted to *Generalmajor* on 1 February 1942. At the end of that year he was made *Inspekteur der Flakschulen und Ausbildungs-Regimenter*, and later *General der Flakausbildung*. All of these posts were in Berlin.

On 1 November 1943 he was given command of 11.Flak-Division, deployed in France, and in February the following year, following his promotion to *Generalleutnant*, he commanded 1.Flak-Division, the anti-aircraft formation assigned the task of defending Berlin, and which came under *Luftgau-Kommando* III. In November 1944 he was posted to this *Luftgau-Kommando* as *General der Flakartillerie*.

Erich Kressmann died on 24 January 1945, near the city of Küstrin, victim of an accident.

Appendix II:
The Equipment of F/88

Flakabteilung 88 received a wide variety of military equipment and armament that was sent out from Germany in order to develop its mission. The most important, and naturally most striking, were the 88mm and 37mm guns and 20mm cannons, although the trucks and tractors, such as the Sd.Kfz. 7 de Krauss-Maffei, Henschel-Diesel, or Krupp Protze also attracted much attention, as did the powerful Siemens 150 cm searchlights and the Elascop Phono-locators, essential for the rapid and efficient detection of enemy warplanes. There were also the generators, optical devices, electro-mechanical Model 36 fire control system, shells, and fuses. All of this materiel provided vital aid to Nationalist air defence capabilities during the course of the Spanish Civil War.

As with the rest of the aviation materiel Germany sent to Spain, the anti-aircraft artillery was the best and most modern available, producing the same structure in the Condor Legion as that found in the *Luftwaffe* at that time (somewhat reinforced), and allowing for the testing in combat conditions of all the magnificent weapons that German industry had been able to design and manufacture in the years preceding the war.

The Germans learned many lessons in the Spanish conflict, but one of the most important for the coming war in Europe was the employment of what would become the most famous artillery weapon of World War II in a ground support role, the ubiquitous 88.

Without the experience gained in Spain, it is very unlikely that the German anti-aircraft artillery would have been as well prepared as it was on 1 September 1939, with the best equipment and the best trained soldiers at that time, the soldiers of the *Heer*.

88/56mm Krupp *Flugzeugeabwherkanone* 18
Designed in 1931, the conditions of the Versailles Treaty made it necessary to give the impression it had been manufactured in 1918 (Flak 18).

Total length in transportation position:	5,800 mm
Length of tube:	4,930 mm
Weight deployed/ towed:	5,150/7,200 kg
Horizontal angle of fire:	360º (two turns each side)
Vertical angle of fire:	from -3º to + 85º
Weight of projectile:	9,240 kg
Muzzle velocity:	830 m/s
Vertical range (effective ceiling):	9,100 m
Horizontal range:	14,800 m
Practical rate of fire:	26 rpm (rounds per minute)

In the Spanish civil war shrapnel shells were used along with AZ/23 and Z.t.S.Z/30 fuses, both percussion and mechanically timed respectively, weighing 9 kg.

The gun had a semi-automatic breech, auto-loader, and automatic fuse adjustment, and was operated in conjunction with the anologic Modelo 36 fire control apparatus.

37/57mm Rheinmetall-Borsig Flugzeugeabwherkanone 18
A piece developed from the Solothurn S-10-100. F-88 was equipped with two troops, each with three guns (one troop in each light battery), mainly assigned to the defence of German aerodromes.

Weight deployed/ towed:	1,748/3,643 kg
Horizontal angle of fire :	360º
Vertical angle of fire:	from -8º to + 85º
Weight of projectile:	0,640 kg
Muzzle velocity:	820 m/s
Vertical range (effective ceiling):	3,523 m
Horizontal range:	6,585 m
Practical rate of fire:	80 rpm (rounds per minute)

Had a semi-automatic sight (the *Flakviser* 33) and used six-round magazines. The gun commander also used a telemeter with a 0.8m base.

20/65mm Rheinmetall-Borsig Flugzeugeabwherkanone 30

F/88 two light batteries were usually organised into four troops, each with three guns, giving the Group a total of twenty-four.

Weight deployed/ towed:	483/770 kg
Horizontal angle of fire:	360°
Vertical angle of fire:	from -12° to + 90°
Weight of projectile:	0,119 kg
Muzzle velocity:	900 m/s
Vertical range (effective ceiling)	2,200 m
Horizontal range:	4,800 m
Practical rate of fire:	120 rpm (rounds per minute)

The guns were mounted on a two-wheeled carriage similar to the 60mm searchlight, and were detached for firing. The weapon mounted a *Flakviser* 30 semiautomatic sight.

The gun commander used a telemeter with a 0.8 m base. The magazines held 20-rounds.

***Komandogerät* (Kdo. Ger.) 36**

Mounted on a Zeiss two-lens telemeter with a four-meter base and using a great many motors, circuits, levers, buttons, and graphs to determine distance and the coordinates of an aircraft, this electromechanical fire control system managed to bring down many enemy planes. It automatically transmitted the data to the receiver needles on all the battery's pieces, allowing the gunners to aim and load the guns. Electric power was supplied to several batteries by a group of generators. The system was operated by a crew of 13 operating the various lenses, telemeter, and the controls for introducing the corrections.

In the absence of the Model 36, the *Raumbdildentfernungsmesser* (Em 4m) telemeter could be employed, mounted on its corresponding pedestal, together with a set of calculation tables and fire grids.

The apparatus weighed 2,200 kg when transported, and it used the same carriage as the 88. Deployed it weighed 1,600 kg.

Krauss-Maffei Sd.Kfz 7 Half-track Tractor

Designed in 1933 as a vehicle capable of towing up to 8 tons and transporting between 15 and 18 men. To tow the twenty 88/56s employed by the five heavy batteries, 20 Sd Kfz 7 type KM-8, KM -9 /10, and KM-11 tractors were used, of which ten remained in Spain at the end of the war.

Engine:	6 cylinder 115/130 hp Maybach HL 52 TU, with 5-speed gearbox
Length:	6,690 mm
Width:	2,350 mm
Height:	2,760 mm
Weight:	11,000 kg
Maximum speed:	50 km/h
Carrying capacity:	1,500 kg

Henschel 33 D1 and G1 6X4 Heavy Truck
Production of the D 1 gasoline engine began in 1934, and the G1 (one of the world's first operational) diesel engine in 1938. In Spain they were used to tow the 88mm and 37mm guns, as well as the Mod. 36 fire control system, searchlights, and the phono-locators.

Engine:	6 cylinder 10.782 cc 100 hp (type D) or 9.123 110 hp (type G) Henschel (six-speed gearbox)
Length:	7,350 mm
Width:	2,250 mm
Height:	2,500 mm
Wheelbase:	3,750 mm
Fording height:	600 mm
Weight (unloaded):	5,000 kg
Carrying capacity:	2,500 kg (3,550 kg towing)
Weight:	4,500 kg
Maximum speed:	55 km/h
Turning radius:	23 m

Krupp L2 H 43 y 143 6x4 Light Truck "Protze"

Production was started in 1933 (the H 43) of a tractor for towing small artillery pieces (in Spain the 20/65); the H 143 appeared in 1937 fitted with a more powerful engine and a different gearbox.

Engine:	4 cylinder 3,308 cc 55 hp (65 hp L 2 H 143) Krupp M 304, with a 5-speed gearbox
Length:	5,100 mm
Width:	19,960 mm
Height:	2,300 mm
Wheelbase:	2,470 mm
Weight:	2,600 kg
Maximum speed:	70 km/h
Carrying capacity:	1,000 kg

150cm *Flak Scheinwerfer* - Siemens MOD. G

The enormous 800 kg Seimens G searchlight, with its one and a half meter diameter mirror set on its platform, was mounted on a carriage similar to the 88. Its maximum range was 13 km and practical range 10 km.

They could be controlled remotely (electrically) by a phono-locator through the corresponding remote-control equipped with telescopes; in the case of being operated independently, the operator would control the searchlight manually using a long bar, with which he could avoid being dazzled by the closeness of the beam. In order to assist in locating the target, the searchlight had a mechanism which allowed the beam to be seperated 9 beams in the form of a cross, although this would then limit range. A 8 kw generator, mounted on a four-wheel trailer, provided energy, and could be positioned up to 200 m from the searchlight.

Elascop Phono-locator
(RRH – *RINGTRICHTERRICHTUNGSHÖRER*- Model 1937)

Manufactured by "Electroacustic" in Kiel, and with two perpendicular 1.36m long microphonic bases, it had a range of between 5 to 12 km, and was theoretically extremely precise. It weighed 723 kg, which increased to 1,000 with its two carriages added. It was equipped with an "Orthogom" adjuster that matched the polar cordinates of the target from the phonolocator to those from the 150 cm searchlight. The searchlights' remote control was a synchronized direction adjuster that transmitted the location of the target (once the parallax had been corrected) detected by the phono-locator to the remote-control (8 augmentations and 8° field of vision) and a 150cm diameter searchlight, whose "visuals" were parallel. The maximum distance between the phono-locator and the telescope/corrector was 50 m, double that between the searchlight and the corrector.

On occasions the data transmitted from two phono-locators was used to better triangulate the position of the target.

Dimensions were length 5.3 m (3.9 m deployed), width 2.11 (1′7 m without running gear), and height 2.9 m (3.1 m with the vertical receivers).

Appendix III: The Flak 14

Although F/88 did not deploy the 7.5 cm L/36 Flak 14, these guns, aquired directly from Berlin by Franco, became the mainstay of Nationalist air defence, becoming the most numerous of the anti-aircraft guns used in the Spanish Civil War.

During the summer of 1937 General Sperrle passed a written offer to Franco's Headquarters, in which the German goverment agreed to provide the Nationalists with

Above and following pages: several images of the German 75/36 mm Flak 14 sold to the National Army in order to complete its Anti-Aircraft Artillery. (J. Negreira)

Alerted in an anti-aircraft battery with a German 75/36 mm. Each man is at his post, ready to fulfill their mission. Enemy aviation will try, by all means, to avoid the area covered by these anti-aircraft guns.

seventeen batteries of 7.5cm and 3.6cm Flak 14s without fire control mechanisms or tractors, although this offer was later increased to twenty-seven batteries. The Artillery Command at Franco's Headquarters assigned Lt Colonel Ayuela the task of travelling to Germany to evaluate the materiel and later report his findings. Following his trip to Germany, Ayuela was concious of deficiencies in the equipment, above all when it was compared to the 88/56 Flak 18, although he still entered a favourable report.

The 7.5 cm. L/36 Flak 14 was a hybrid, a product of the adaptation of the tube and cradle of a field gun manufactured by Rheinmetall (designated the FK 16 nA, which had then entered service in 1934) to the cruciform base of the Krupp 7.5 cm. L/60 prototype. The *Feldkanone* FK 16 nA itself was the result of the coupling of a 7.5cm tube to the carriage of the First World War vintage 7.7cm FK 16, which gave rise to the abbreviation nA (*neuer Artillerie*). This piece had been a temporary solution, developed at the beginning of the thirties at the insistance of the *Heer*, while Krupp were finalizing the design of the 8.8 cm. Flak 18.

The 7.5 cm. L/36 Flak 14 never came to equip units in Germany, and it is highly likely that all the units produced were sold to the Spanish Nationalists.

In the end the Artillery General Headquarters approved the order, although the number of batteries that arrived did not coincide with the original offers, as twenty-one batteries were delivered in total, giving a total of ninety Flak 14s (eighty-four deployed and six in reserve). The initial German offer, which excluded transport for the batteries, was rejected by the Artillery General Headquarters. Ayuela insisted they were necessary, at least five tractors per battery, with Germany finally agreeing to supply one hundred and five six-wheel Henschel trucks.

What is certain is that the 7.5 cm. L/36 Flak 14 had its limitations, and not having entered service at home also went against it. To this can be added the fact that there were certain technical aspects which made it less favourable as an anti-aircraft gun when compared to contemporary models in the service of other countries, as well as its lack of a fire control system.

However, in spite of this the ninety Flak 14s that arrived in Valladolid at the beginning of 1937 were to form the backbone of the Nationalist anti-aircraft artillery. In December that same year they were handed over to the artillery units that had been selected to operate them, and in February 1938 they were integrated into the Anti-Aircraft Artillery Group, and numbered batteries 21 to 42. In June 1938 they became part of the then recently organized National Army Anti-Aircraft Regiment.

In spite of their limited use in the Anti-Aircraft role, these pieces were in service with the Spanish Army for more than two decades, and were still on the Artillery's inventory in 1963.

Previous page: the Lerida Front, 1938. Several commanders and artillery officers pose in front of a 75/36 mm Flak 14. (Authors)

Above: the Parade in Barajas; German motorcycles and 75/36 mm pieces. (C. Franco)

Appendix IV: F/88 and the Farewell Parades of the Condor Legion

Following the end of hostilities and its departure for Germany, F/88 took part in all the victory parades held throughout Spain, with a representation consisting of both its personnel and equipment.

Leon, Barajas, Zaragoza, and the great victory parade celebrated in the Paseo de la Castellana, in Madrid, were perhaps the most important acts in which the *Luftwaffe's* Anti-Aircraft unit participated in Spain, parading along the streets with their artillery pieces in front of the authorities and an ecstatic public, eager to celebrate the return of peace after almost three long years of war.

The parade in Barajas (5-12-1939); aircraft in the background and personnel passing by in front of the platform, *Flakabteilung* 88. (C. Azaola)

Barajas, 12 May 1939. The indisputable stars of the parade were the 88/56 Flak 18s, towed by the imposing 8-ton Krauss-Maffei tractors (CECAF)

The Flak 18's younger brothers were also represented in the Barajas parade. Here they can be seen being towed by the sensational Krupp "Protze" H-2 L43 light tractors. (CECAF)

The great victory parade held in Madrid on 19 May 1939. The Condor Legion naturally participated with both its foot and motorized units. The photographs show F/88. (Via the Authors)

Appendix V: List of the Fallen

-BAUER, Friedrich	*Unteroffizier*	+ 6-12-37 in Morga (Vizcaya) shot after mistakenly entering enemy lines, together with three comrades.
-BAUER, Wolfgang	*Unteroffizier*	+ 7-6-38 in Castellon, according to some reports he was shot by an enemy firing squad, other sources claim he died alter being hit in the head by an incoming 20 mm round. He belonged to 5. F/88 and had arrived in Spain the previous February.
-BODE, Siegfried	*Obergefreiter*	+ 4-9-38 following a motorcycle accident in the vicinity of Alcarraz (Lerıda).
-CLAUS, Felix	*Unteroffizier*	+ 6-12-37 en Morga (Vizcaya) shot after mistakenly entering enemy lines, together with three comrades.
-COBURGER, Helmut	*Obergefreiter*	+ 2-19-39 fell to his death in Sabadell (Barcelona).
-CONRAD, Richard	*Obergefreiter*	+ 25-12-38 in Pineda de la Sierra (Burgos) alter a Ju 52 of 2. K/88 overturned carrying a cargo of shell fuses destined for the Catalan Front. Another four men were also killed in the incident and three survived.
-CREUTZ, Emil	*Gefreiter*	+ 4-4-37 in Hospital in Vitoria from wounds suffered when a shell accidentally exploded near Urbina (Alava).

-EPPERT, Rudolf	*Gefreiter*	+ 1-14-37 in Villanueva del Pardillo (Madrid) after being shot in the chest.
-FINGER, Heinrich	*Gefreiter*	+ 12-30-37 in Teruel from a head wound.
-FISCHER, Johann	*Gefreiter*	+ 4-20-37 in Hospital in Vitoria from wounds suffered when a shell accidentally exploded near Urbina (Alava).
-FLOREZACK, Ludwig	*Unteroffizier*	+ 12-28-37 aerial bombing in Bezas (Teruel)
-GAUS, Gustav	*Oberfeldwebel*	+ 30-4-37 when the mail plane on the route Seville-Rome in which he was traveling disappeared. According to some witnesses the aircraft crashed into the sea near the port of Malaga.
-HENKE, Werner	*Unteroffizier*	+ 8-23-38 in a traffic accident in San Rafael del Río (Castellon)
-HENSEN, Otto	*Unteroffizier*	+ 1-17-37 when the mail plane he was aboard crashed in the Sierra de Gredos, at a point known as "El Torreon", located at the dividing line between Jerte (Caceres) and Candelario (Salamanca).
-HILGER, Gerhard	*Gefreiter*	+ 10-4-37 in Griñon (Madrid) from pneumonia.
-HOFFMANN, Martin	*Gefreiter*	+ 12-6-37 en Morga (Vizcaya) shot after mistakenly entering enemy lines, together with three comrades.
-HORN, Ehrhardt	*Gefreiter*	+ 26-6-38 in Villarreal (Castellon) from enemy shellfire ahainst his position. Member of 5. F/88.

-HÜSEMANN, Heinrich	*Obergefreiter*	+ 2-4-39 shot by firing squad in Roda de Ter (Gerona).
-KNÖHR, H.-Joachim	*Gefreiter*	+ 11-1-39 when the piece he was serving received a direct hit during an engagement with enemy tanks near Espluga de Francoli (Tarragona). Member of 4. F/88.
-KOHLHEIM, Georg	*Gefreiter*	+ 11-9-37 killed by shrapnel in Llanes (Asturias).
-LAMPE, Arno	*Unteroffizier*	+ 5-1-38 killed by explosion in Teruel. Member of 6. F/88.
-MÜLLER, Walter	*Kanonier*	+ 2-7-38 in Bechi (Castellon). Probably shot by the enemy.
-OTTO, Hermann	*Gefreiter*	+ 15-8-37 in Alar del Rey (Palencia) killed by a Spanish soldier.
-PLEWE, Erwin	*Obergefreiter*	+ 11-1-39 when the piece he was serving received a direct hit during an engagement with enemy tanks near Espluga de Francoli (Tarragona). Member of 4. F/88.
-REMKE, Heinrich	*Obergefreiter*	+ 28-3-39 in Orgaz (Toledo) when a ammunition vehicle exploded. Member of 2. F/88 and had arrived in Spain in March 1938.
-RETTENMAIER, Karl	*Gefreiter*	+ 4-22-37 in Hospital in Vitoria from wounds suffered when a shell acidentally exploded near Urbina (Alava).
-SCHÄFER, Konrad	*Gefreiter*	+ 12-24-36 from a hand-grenade explosion in the province of Madrid.
-SIMON, Alfred	*Obergefreiter*	+ 6-5-38 from a hand-grenade explosion in Benasal (Castellon). Member of 4. F/88.

-SORDON, Georg	*Obergefreiter*	+ 1-12-39 when the piece he was serving received a direct hit during an engagement with enemy tanks near Espluga de Francoli (Tarragona) the previous day. Member of 4. F/88.
-STEEG, Richard	*Unteroffizier*	+ 6-12-37 in Morga (Vizcaya) shot after mistakenly entering enemy lines, together with three comrades.
-VANDREY, Erich	*Feldwebel*	+ 7-8-38 in Castellon from enemy shellfire while serving his artillery piece some days earlier. Member of 5. F/88.
-WOLF, Felix	*Dolmetscher*	+ 14-7-37 automobile accident in Santo Domingo (Vizcaya).
-ZSCHETZSCHING, Erich	*Gefreiter*	+ 15-4-37 when a grenade he was handling exploded, in Santa Agueda, Mondragon (Guipuzcoa).

Bibliography

Chamberlain, Peter and Terry Gander: *88 FLAK and PAK: A Profile Special*, New York, Arco Publishing, 1976.

Fleischer, Wolfgang: *Die 2-cm Flugzeugabwehrkanonen 30 und 38*, Wölfersheim, Podzun-Pallas, 2002.

Griehl, Manfred: *Das Grosse Buch der Flak. Deutsche Luftverteidigung 1912-1945*, Wölfersheim, Podzun-Pallas, 2003.

Grimme, Hugo: *Ehrenblatter der deutchen Flakwaffe*, Berlin, Bernard & Graefe, 1940.

Gmeline, Patrick de: *"La Flak:" 1935-1945: la DCA allemande*, Bayeux, Heimdal, 1986.

Koch, Horst Adalbert: *Flak: Die Geschichte der deutschen Flakartillerie*, Bad Nauheim, Podzun, 1954.

Koch, Horst Adalbert y Schindler, Heinz: *Flak: Die Geschichte der deutschen Flakartillerie und der Einsatz der Luftwaffenhelfer*, Bad Nauheim, Podzun-Pallas, 1965.

Manrique García, Jose Mª; Molina Franco, Lucas; Mortera Perez, Artemio: *Historia de la Artilleria Antiaérea Española, volumen I*, Valladolid, Quirón ediciones, 1998.

Molina Franco, Lucas: *El legendario cañón antiaéreo de 88mm. Su historia y evolucion en el Ejercito español*, Valladolid, Quirón ediciones, 1996.

Müller, Werner: *German 20mm Flak in World War II, 1934-1945*, Atglen, PA, Schiffer, 1995.

Müller, Werner: *The heavy Flak guns 1933-1945*, Atglen, PA, Schiffer, 1990.

Müller, Werner: *The 88mm Flak in the First and Second World Wars*, Atglen, PA, Schiffer, 1998.

Müller, Werner: *The 88mm Flak* , Atglen PA, Schiffer , 1991.

Müller, Werner: *The Schwere Flak, Utting*, Nebel Verlag, 2000.

Neuman, Ernst: *Handbuch fur den Flakartilleristen: Der Kanonier, Waffen und Ausbildung der schwere Flakbatterie- 8.8 cm-Flak, 18 und 2 cm-Flak 30*, Berlin, "Offene Worte", 1938.

Norris, John : *88mm Flak 18/36/37/41 & PaK 43 1936-45*, (col. New Vanguard nº 46), Oxford, Osprey, 2002.

Piekalkiewicz, Janusz: *Die 8.8 Flak im Erdkampf-Einsatz*, Stuttgart, Motorbuch Verlag, 1978.

Piekalkiewicz, Janusz: *The German 88 Gun in Combat. The Scourge of Allied Armor*, Atglen, PA, Schiffer, 1992.

Trojca Waldemar: *8,8 cm Flak 18-36-37*, Zweibrücken, VDM Heinz Nickel, 2003.

Westermann, Edward B.: *Flak: German Anti-Aircraft Defenses, 1914-1945, (col. Modern War Studies nº 5)*, Lawrence, Kansas, University Press of Kansas, 2001.

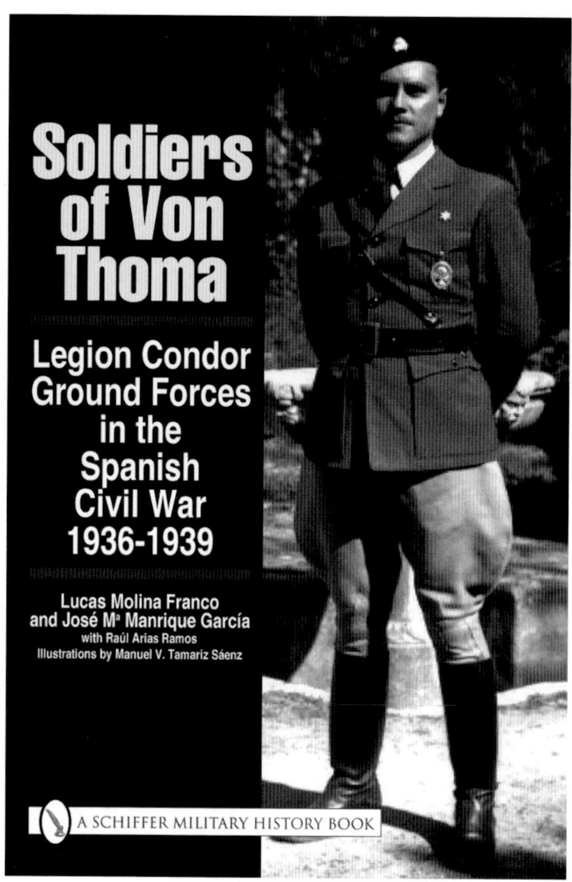

Soldiers of von Thoma
Legion Condor Ground Forces
in the Spanish Civil War

Lucas Molina Franco & José Ma Arias Ramos

In October 1936 two German ships arrived in Spain loaded with a Panzer I company and all the services and personnel to establish Franco's army's first armored unit, the so-called Panzer-Gruppe Drohne. The Third Reich chose a pioneer in the development of German ground forces to command this contingent - Oberstleutnant Wilhelm Ritter von Thoma. This book presents the history of von Thoma's units in the Spanish Civil War, from the tank crews to the infantry, and specialties.

ISBN: 9780764329265, 6"x9",
over 200 bw/color images, 240 pp., hard cover
$39.95

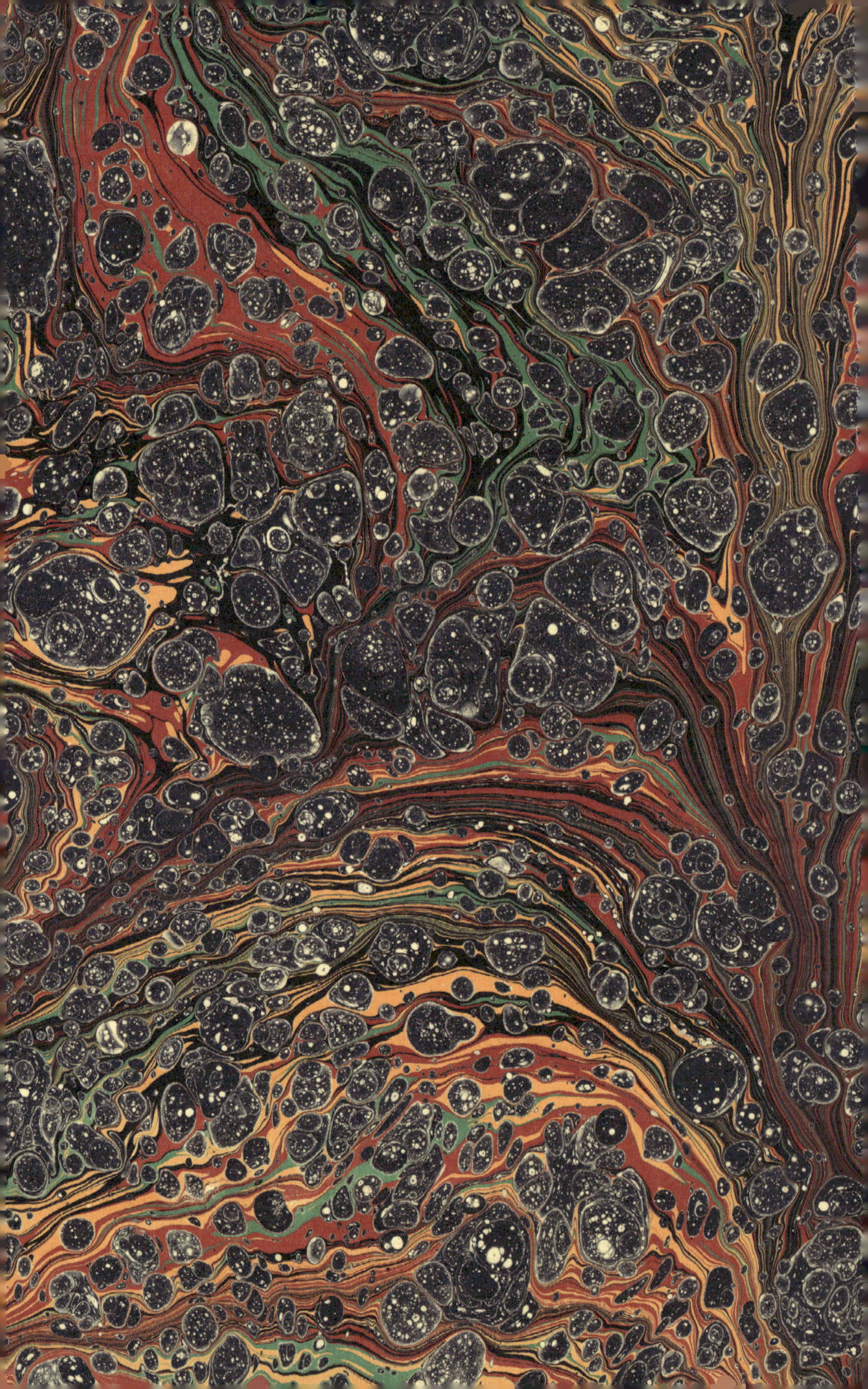